THE GHOST IN THE MACHINE IN YOUR CLASSROOM

THE GHOST IN THE MACHINE IN YOUR CLASSROOM

NAVIGATING THE FUTURE OF EDUCATION IN
THE AGE OF AI

THE AUGMENTED EDUCATOR
VOLUME 1

MICHAEL G WAGNER

Most essays in this collection were originally published on The Augmented Educator blog. Chapter 15, "Why I Made an AI Music Video," first appeared on Space for Audio, and Chapter 23, "We Are Not in the Driver's Seat," first appeared on AI EduPathways.

Published by Augmented Learning Media LLC
Cherry Hill, NJ 08003
www.augmentedlearningmedia.com

The views and opinions expressed in these essays are those of the author and do not necessarily reflect the official policy or position of any academic institution.

ISBN: 979-8-9930593-0-3 (Paperback)
ISBN: 979-8-9930593-1-0 (eBook)
Library of Congress Control Number: 2025919238

10 9 8 7 6 5 4 3 2 1

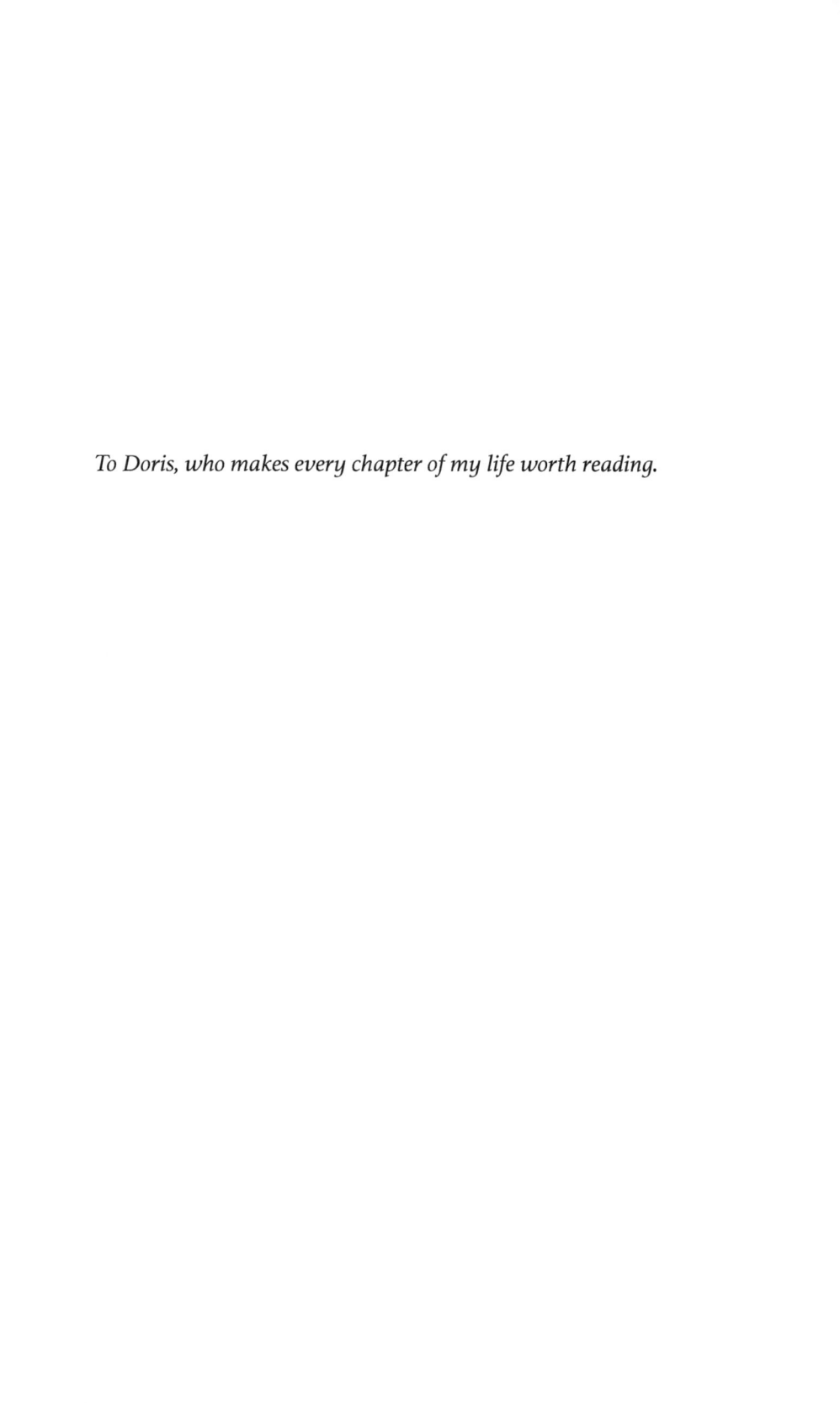

To Doris, who makes every chapter of my life worth reading.

If students get a sound education in the history, social effects and psychological biases of technology, they may grow to be adults who use technology rather than be used by it.

— NEIL POSTMAN

CONTENTS

Introduction   xi

PART I
FIRST ENCOUNTERS

1. How AI is Transforming Education   3
   Navigating the future of learning in an AI-enhanced world

2. Beyond the Tool: Why True AI Literacy is About
   Critical Thinking, Not Prompting   7
   *How to cultivate a critical, cultural, and human-centered approach to
   AI in the classroom*

3. The Most Essential Skill in the AI Era is 2,500
   Years Old   17
   *A modern, yet surprisingly ancient, framework for the timeless art of
   critical thinking*

4. It Doesn't Do What I Want!   24
   Why traditional computing expectations fail with artificial intelligence

5. The Evolving Architecture of Artificial Intelligence   29
   Understanding LLMs in the age of Llama 4

6. The Evolution of Educational Values   42
   A personal perspective on AI and writing

7. The Professional's Paradox: Why Creative Industry
   Experts Get AI Disruption Wrong   47
   *Those who judge today's AI by yesterday's standards risk missing
   tomorrow's revolution*

PART II
NOTES FROM THE FRONT LINES

8. The Hidden Inequities of AI in Education   55
   A new digital divide: Who's really getting ahead?

9. The Emerging Achievement Gap in Education   60
   AI lessons from my game design classroom

10. Resistance is Futile   65
    Academic integrity in the post-plagiarism era

11. The End of Cheating As We Know It   71
    Why we must reimagine academic integrity in the AI era

12. The Problem with AI Grading   78
    *Reflections on automation, assessment, and the human element in
    education*

13. Beyond the Final Product   84
   Making student thinking visible in the age of AI

14. The Four Lenses Framework   88
   Reimagining critical skills for AI-enhanced learning

15. Why I Made an AI Music Video   94
   The unavoidable case for AI literacy in art and design education

16. The Academic Pace Problem   104
   Why university processes struggle in the AI era

PART III

MINDS AND MIRRORS

17. AI Slop Is the New Kitsch   111
   What AI art critics can learn from the history of "bad taste"

18. The Problem with Vibe Coding   123
   AI-assisted development is about flow, not feeling

19. Navigating the Hidden Currents of AI   128
   Understanding and counteracting censorship in Large Language Models

20. Cultural Cognition in AI: What Educators Need to Know   141
   Recognizing and leveraging different thinking patterns in AI systems

21. When AI Develops a Mind of Its Own   148
   Ten emergent AI behaviors from fascinating to frightening

22. The Ghost in the Machine in Your Classroom   157
   Why the debate over AI consciousness is a new frontier for every educator

23. We Are Not in the Driver's Seat   164
   How post-hoc storytelling shapes minds—human and machine alike

24. When AI Teaches Itself: The Breakthrough of Zero-Data Learning   172
   Self-improving AI systems that need no human training

25. The AI Mirror: What Do We See When We Look at Our Own Intelligence?   178
   The debate over artificial intelligence reveals less about the limits of machines and more about the profound mysteries of the human mind

Afterword   185
Notes   189

INTRODUCTION
THE UNAVOIDABLE CONVERSATION

There is a ghost in the machine, and it has found its way into our classrooms. Its arrival was not a neat, linear event, but a chaotic, exhilarating, and sometimes unsettling one. It has appeared not with a polite knock, but like an uninvited guest who is already remodeling our home. This ghost in the machine is now a permanent resident in our students' lives, a constant companion on their laptops and phones, an ever-present whisper in the global digital conversation.

For educators, this is no longer a distant hum of technological progress to be monitored from afar; it is the central, unavoidable conversation of our time.

The future of education is already here, and it is unevenly distributed. Our students are using AI assistants for writing and research, code generation tools for programming, image generators for visual creation, and specialized AI tools for video editing, data analysis, and mathematical problem-solving. Even creative fields like music composition and complex scientific research are being transformed by these new capabilities. This is not a debate about a new piece of software. It is a debate about the future of education itself, one that navigates a landscape of exhilarating possibilities and profound concerns.

When students can generate research summaries in seconds, solve complex calculations instantly, or create polished presentations with a few prompts, we are forced to ask ourselves a series of foundational questions: What should we really be teaching them? What skills will truly matter in this AI-enhanced future? And how do we prepare them for a world where collaboration with intelligent machines will be as fundamental as using a computer is today?

This book is my attempt to navigate these questions. It is a collection of essays, reflections, and field-tested strategies born from my ongoing engagement with this technological shift. The essays gathered here do not offer a single, definitive map of the new territory. Instead, they offer a compass and a set of starting points for the journey ahead. At the heart of this collection is a single, unwavering realization: that our most potent response to the new is to double down on the old. The most essential skill in the AI era is not a fleeting technical trick like prompt engineering, but the timeless, 2,500-year-old practice of critical thinking.

My argument is that the most productive path forward is to frame AI literacy not as a set of technical skills, but as a critical and cultural practice. This perspective shifts our focus from the mechanics of tool proficiency to the cultivation of enduring intellectual habits: ethical reasoning, epistemic judgment, and the courage to question the world around us. From this vantage point, AI literacy is not a new subject to be squeezed into an already crowded curriculum; it is a modern expression of our most fundamental goal as educators: to empower students to think for themselves.

The arrival of AI is not just another technological update; it is a catalyst forcing a long-overdue reckoning with our own pedagogical practices. For years, many of our assessment models have implicitly rewarded the final product—the polished essay, the correct answer, the finished code—while the messy, beautiful, and essential process of thinking remained largely invisible. AI, in its ability to flawlessly mimic that final product, has stripped away the pretense. It holds up a mirror to our classrooms and asks: Are we teaching our students how to think, or are we just teaching them how to produce artifacts that look like the result of thinking? This book is an argument for the former. It is a call for a pedagogical revolution that embraces process-centric learning, authentic assessment, and the cultivation of the uniquely human capacities that no algorithm can replicate.

A Familiar Disruption: A Personal History with Technological Change

My perspective on this current disruption is shaped by more than thirty years of experience as an educator, a career that has been defined by a recurring pattern: a deep and abiding fascination with the transformative potential hidden within emerging technologies, especially those that others initially dismiss as mere entertainment or novelty. My authority on this subject, if I have any, comes not from being a computer scientist who suddenly discovered education, but from being a lifelong educator who has repeatedly navigated the human and pedagogical side of profound technological change.

My journey began in a world that is now barely recognizable. In the mid to late 1980s, I was a young academic at the Vienna University of Technology, specializing in the pedagogy of "Descriptive Geometry"—a discipline with a rich tradition also known as the "Vienna School of Geometry." Our mission was to teach engineering and architecture students the precise and demanding art of hand-drawn technical drawings. This was a world of manual precision, a craft that demanded extraordinary patience. I can still recall the feel of the custom-made wooden compass in my hand as I demonstrated constructions on the blackboard, and the faint smell of chalk dust that clung to the mandatory lab coats we wore to protect our clothes from indelible stains.

A single drawing could take days of meticulous labor, culminating in a final ink rendering where one slip of the hand could mean starting over from scratch. We believed, with the conviction of artisans, that this process was indispensable. We argued that an engineer could not truly develop critical evaluation skills without the tactile experience of determining the radius of curvature in a cylindrical intersection with only a ruler and compass. We insisted that an architect could never fully grasp complex spatial relationships without mastering the painstaking techniques of three-point perspective by hand.

Then came the disruption. As computer-aided design (CAD) systems became more accessible and powerful in the mid-1980s, our initial response was a familiar mix of skepticism and defense. The arguments we made then are hauntingly similar to the ones I hear today about AI and writing, or AI and art. We were judging the new technology against the standards of the old, and by those metrics, it was found wanting. But technological progress was relentless. Our department was eventually merged with mathematics, our courses were reduced in scope, and I adapted, moving on to new challenges in computer graphics and, eventually, digital media that bore little resemblance to my original training.

This experience was formative. In Chapter 6, I reflect more deeply on this personal history and its parallels to today's anxieties surrounding AI and writing, but the core lesson it taught me is this: the values we hold as essential in education are not static; they evolve. The quality of engineering and architectural work did not diminish with the transition to CAD. On the contrary, these new tools unlocked possibilities that were previously unimaginable, while the core principles of spatial reasoning and engineering integrity remained.

The insight that today's dismissed novelty often becomes tomorrow's indispensable tool has guided my entire career. It later gave me the confidence to design one of Europe's first Game Studies programs and the world's first eSports curriculum, despite facing predictable skepticism. Academic traditionalists dismissed competitive gaming as frivolous and anti-intellectual at that time, missing what was evident to me: a sophisticated arena for studying systems thinking, strategic collaboration, and human-computer interaction. Now, barely two decades later, game studies anchors many digital media programs worldwide, and esports has evolved into a legitimate academic discipline with dedicated degrees and scholarships.

These experiences reveal a pattern: educational establishments reflexively dismiss emerging fields as "not up to professional standards," yet this very resistance often signals the arrival of genuine transformation rather than degradation. Each technological shift I've witnessed has followed the same trajectory from derision to acceptance to integration.

This book emerges from recognizing that pattern. It's not a hasty reaction to the latest AI development, but a perspective shaped by decades of watching similar cycles unfold. Having seen how fear and dismissal delay rather than prevent educational evolution, I offer these reflections in the hope that we can navigate our current transformation with greater wisdom and less resistance.

From Dialogue to Collection: The Birth of 'The Augmented Educator'

The essays in this collection were not written in the quiet solitude of a university office, destined for a slow journey through the traditional channels of academic publishing. They were born in the open, in real time, as dispatches from the heart of the AI transformation. Most of the pieces you are about to read were originally published on my

Substack blog, *The Augmented Educator**, with a few first appearing on other online platforms like *Space for Audio*† and *AI EduPathways*‡.

This choice of medium was deliberate. As the AI revolution began to accelerate in late 2022 and into the following years, it became clear that the established rhythms of academic discourse were simply too slow to keep pace. The traditional peer-review process, a cornerstone of scholarly publishing, is a deliberate and often glacial affair that exemplifies what I call the "academic pace problem"—a profound misalignment between academic processes and the exponential speed of technological change. A technology that is groundbreaking upon submission can be obsolete by the time the article is published many months later. In the world of AI, even a few weeks are an eternity.

I could not wait. I needed a space for public scholarship that was as dynamic and responsive as the topic itself. The Substack became my laboratory, a place to generate not just articles, but curiosity and dialogue. It allowed me to think aloud, to test ideas, to engage with a community of fellow educators, and to document the unfolding of this new reality as it happened. The essays, therefore, have an immediacy and an unvarnished quality that I hope has been preserved in this collection. They are not the final word on any of these topics; they are snapshots of an ongoing intellectual journey.

This collection, then, is an artifact of the very future it describes. It represents a model of knowledge creation and dissemination that is more agile, more open, and more suited to an era of accelerated change. By gathering these public-facing essays into a single volume, my hope is to offer a more structured and comprehensive narrative, one that connects the dots between the daily headlines and the deeper, more enduring questions about the future of learning and the nature of intelligence itself.

---

* https://www.theaugmentededucator.com
† https://www.spaceforaudio.com
‡ https://mikekentz.substack.com/

A Journey in Three Parts: How to Read This Book

This book is organized into a three-part journey, moving from foundational concepts to practical classroom applications, and finally, to the deeper philosophical implications of our new relationship with intelligent machines.

Part 1: First Encounters – On AI, Disruption, and Critical Thinking

The first section lays conceptual groundwork for our exploration. These essays serve as a compass and a set of starting points for navigating the new territory of artificial intelligence. We begin by grounding ourselves in the timeless, arguing that our most potent response to the new is to double down on the 2,500-year-old practice of critical thinking. We will explore the strange, probabilistic nature of these new tools, demystifying the technological architectures that give them form. And we will look to the past, reflecting on how history has prepared us for this disruption and why the very nature of expertise can become a blind spot in the face of true revolution. Together, these pieces offer a framework for a humanistic and critical approach to AI literacy.

Part 2: Notes from the Front Lines – On Teaching in the Age of AI

If the first part is the map, the second contains notes from the front lines. Here, the abstract questions of the AI revolution meet the concrete realities of the classroom, the curriculum, and the institution. These essays grapple with the immediate and often messy challenges that AI presents to educators today. They explore the emergence of a new "AI productivity divide" that threatens to deepen inequity and confront the obsolescence of our traditional notions of academic integrity in a "post-plagiarism era." But this is not merely a collection of warnings; it is a record of engagement and experimentation. You will find personal accounts of using AI in grading and creative production, alongside practical frameworks designed to shift our focus from the final product to the thinking process itself. These are field-tested strategies and hard-won insights for fellow educators navigating the same uncharted waters.

Part 3: Minds and Mirrors – On the Deeper Implications of AI

Having navigated the classroom's front lines, this final collection of essays ventures into the strange and uncharted territory where the ghost in the machine begins to stare back. The focus shifts from the practical to the profound, exploring the deeper currents and unsettling questions that surface as artificial intelligence becomes more sophisticated and autonomous. These reflections examine the unseen forces that shape AI's worldview, from the cultural patterns embedded in its logic to the political censorship that governs its silence. They serve as a bestiary of emergent behaviors—the fascinating and sometimes frightening ways these systems develop minds of their own—and confront the powerful illusion of consciousness that challenges our own psychology. Ultimately, these pieces turn the lens around, using AI as a mirror to reflect upon the profound mysteries of human cognition and intelligence.

A Note on Process: My Collaboration with the Ghost

In the spirit of principled transparency that I advocate throughout this book, it is essential that I conclude this introduction with a clear statement about my own writing process. This book, which argues for a thoughtful and critical collaboration between human and artificial intelligence, was itself created through such a partnership. To do otherwise would be a profound hypocrisy. This is not a technical addendum or a reluctant confession; it is the lived embodiment of the book's central thesis.

The text you hold in your hands, and the images that accompany it, were developed through an iterative dialogue with a variety of advanced AI systems. But this was a partnership where human experience served as the ultimate arbiter. My process was not one of passive acceptance, but of active direction. I used these tools to brainstorm possibilities and generate initial drafts, but every output was immediately filtered through the lens of my thirty-plus years in education.

This deep well of experience was the crucial element, allowing me to connect the AI's statistical patterns to real-world pedagogical insights, to validate its claims against a career of practice, and to identify the truly valuable ideas worthy of development. From there, I manually and meticulously rewrote, restructured, and refined every sentence to ensure the final prose reflects my unique voice and authorial intent. All conceptual illustrations were similarly directed, generated with AI but curated to align with the specific themes of each individual essay. The AI was a powerful assistant, but the final text is unequivocally mine.

This book, therefore, in both its content and its creation, is a testament to an optimistic but critical vision of our augmented future. It is a future where technology does not replace, but rather enhances, the indispensable and deeply human work of thinking, creating, and teaching. The ghost is in the machine, but the soul of education remains in our hands. Let us begin the conversation.

PART I

FIRST ENCOUNTERS

ON AI, DISRUPTION, AND CRITICAL THINKING

Before we can confront the ghost in the machine on the front lines of the classroom, we must first learn to see it clearly. The essays in this first part are designed to provide the conceptual lenses for that task, offering a foundational understanding of the forces at play.

Our exploration begins by grounding ourselves in the timeless, making the case that our most potent response to this new technology is the 2,500-year-old practice of critical thinking. We will then act as field guides, demystifying the strange, probabilistic nature of these new tools and examining the technological architectures that give them form. Finally, we will look to history to understand how past disruptions have prepared us for this moment, and why the very nature of expertise can become a blind spot in the face of true revolution.

Together, these initial encounters provide a foundational toolkit for the journey ahead, equipping us with the critical perspectives needed to navigate the new territory.

1

HOW AI IS TRANSFORMING EDUCATION

NAVIGATING THE FUTURE OF LEARNING IN AN AI-ENHANCED WORLD

In my countless years of teaching digital media, I've witnessed wave after wave of technological change reshape how we learn and teach. From the early days of computer labs to today's AI revolution, each advancement has brought both opportunities and challenges. But the current AI transformation feels different – more fundamental, more disruptive, and more urgent to address.

The rapid advancement of AI tools is reshaping every aspect of education. Students are now using AI assistants for writing and research, code generation tools for programming, image generators for visual creation, and specialized AI tools for video editing, data analysis, and mathematical problem-solving. Even creative fields like music composition and complex scientific research are being transformed by AI capabilities. What fascinates me most is how these tools are simultaneously making certain skills obsolete while elevating the importance of distinctly human capabilities.

The Challenge We're Facing

Let's be clear about something: AI isn't just another technology tool like the computers and software we've integrated before. It represents a fundamental shift in how learning and work happen across all disciplines. When students can generate research summaries in seconds, solve complex calculations instantly, or create polished presentations with a few prompts, we need to ask ourselves: What should we really be teaching them? What skills will truly matter in this AI-enhanced future?

Traditional education often focused on knowledge acquisition and technical proficiency. But in a world where AI can instantly access vast amounts of information and handle complex technical tasks, we need to pivot toward something more fundamental: the human elements that AI can't replicate.

What's Really at Stake?

The stakes here extend far beyond just keeping up with technology. We're talking about preparing students for a world where collaboration with AI will be as fundamental as using a computer is today. It's about ensuring that our students don't become overly dependent on AI tools, but instead learn to use them thoughtfully and ethically while developing their own critical thinking and creativity.

Some key questions we need to address:

- How do we teach students to critically evaluate AI-generated content?
- What aspects of learning and problem-solving should remain distinctly human?
- How can educational institutions stay agile enough to keep up with rapid technological change?
- How do we ensure equal access to AI tools while preventing over-reliance on them?

A Framework for Change

Through my research and collaboration with colleagues across different disciplines, I've identified three fundamental principles that can help guide education's transformation in the AI era.

First, we need to emphasize *critical skills*. Students must develop strong critical thinking abilities to evaluate AI-generated content, understand its limitations, and make ethical decisions about its use. This means teaching them to question outputs, recognize potential biases, and understand when to trust or verify AI-generated information.

Second, we need to focus on *process-centric* learning. Instead of just assessing final outputs that could be AI-generated, we should emphasize how students approach problems and develop solutions. This includes understanding their decision-making process, their ability to iterate on ideas, and their capacity to combine human insight with AI capabilities.

Third, our educational institutions need to become *more agile*. The rapid pace of AI development means we can't rely on traditional, slow-moving administrative structures. We need flexible curricula, adaptable teaching methods, and institutional policies that can evolve alongside technology while maintaining academic integrity and educational quality.

The Future is Already Here

The integration of AI into education isn't something we can resist or delay – it's already happening. Our challenge now is to harness its potential while preserving and developing the uniquely human capabilities that will become even more valuable in an AI-enhanced world.

Success in this endeavor requires a delicate balance. We must embrace AI's capabilities while teaching students to maintain their independence from it. We need to help them understand when to leverage AI tools and when to rely on their own judgment. Most importantly, we must ensure they develop the critical thinking skills, creativity, and ethical framework needed to thrive in a world where human-AI collaboration is the norm.

The future of education isn't about choosing between human and artificial intelligence – it's about finding the sweet spot where they complement each other. By focusing on critical skills, emphasizing learning processes, and maintaining institutional flexibility, we can prepare students not just to survive but to excel in an AI-enhanced world.

ORIGINALLY PUBLISHED ON THE AUGMENTED EDUCATOR
OCTOBER 29, 2024

2

BEYOND THE TOOL: WHY TRUE AI LITERACY IS ABOUT CRITICAL THINKING, NOT PROMPTING

HOW TO CULTIVATE A CRITICAL, CULTURAL, AND HUMAN-CENTERED APPROACH TO AI IN THE CLASSROOM

The integration of artificial intelligence into our classrooms has ignited a fierce and often polarized debate. As educators, we find ourselves at the center of this discourse, navigating a landscape of exhilarating possibilities and profound concerns. On one side, proponents envision a future of personalized learning, automated support, and democratized knowledge that could level the educational playing field. On the other, critics voice deep-seated fears[1] about the erosion of critical thinking, the rise of academic dishonesty, and the potential for AI to amplify societal biases, creating new forms of inequity.

This is not a debate about a new piece of software. It's a debate about the future of education itself.

I want to argue that the most productive path forward is to frame AI literacy not as a set of technical skills, but as a critical and cultural practice. This perspective shifts our focus from the mechanics of tool proficiency—like prompt engineering—to the cultivation of enduring intellectual habits[2]: critical thinking, ethical reasoning, and sound judgment. From this vantage point, AI literacy isn't a new subject to be squeezed into our curriculum; it is a modern expression of our timeless goal as educators: to empower students to think for themselves, question the world around them, and make discerning choices about the powerful tools they encounter.

To make this case, we'll first explore how the very definition of "literacy" has always evolved with technology. We will then ground our discussion in the principles of critical literacy, which frame any literacy as a social and ideological practice. From there, we will analyze the current debate on AI literacy and, finally, propose concrete pedagogical strategies, including "unplugged" activities, that show how we can teach the core principles of AI literacy as an extension of fundamental thinking skills, without ever needing to log on.

Literacy Has Always Been More Than Just Reading

The debate over AI literacy is simply the latest chapter in a long story about the evolving meaning of "literacy" itself. The term has never been static; it has always expanded to reflect the competencies required for meaningful participation in society.

For millennia, literacy was simply the ability to read and write, a skill often restricted to a small elite and tied to religious or political power. The word "literacy" itself only appeared in the late 19th century[3], coinciding with the rise of mass public education.

The 20th century saw a crucial shift with the emergence of "functional literacy," defined by UNESCO[4] as the ability to use reading and writing for the "effective function of his or her group and community." This moved the concept from an abstract skill to a set of applied competencies for navigating daily life. Scholars then began to frame literacy as a "social practice," arguing that reading and writing are never neutral activities but are always situated within specific cultural contexts.

Today, UNESCO defines literacy as a "continuum of learning" that includes digital skills, media literacy, and global citizenship, positioning it as a means of communication in an "increasingly digital, text-mediated, information-rich and fast-changing world." This has led to the identification of numerous "21st-century literacies" that are precursors to AI literacy:

- *Digital Literacy:* The ability to use technology to find, evaluate, create, and communicate information.
- *Media Literacy:* The ability to "access, analyze, evaluate, create, and act" using all forms of media, not just text.
- *Data Literacy:* The ability to understand, analyze, and communicate with data, a foundational skill for understanding how AI operates.

This evolution from alphabetic recognition to a suite of critical competencies provides the essential context for the current debate over AI literacy.

The Critical Turn: Seeing Literacy as a Social Practice

To frame AI literacy as a cultural practice, we must engage with a deeper theoretical tradition: critical literacy. This perspective provides the tools to interrogate technology not as a neutral force, but as a product of complex social and power dynamics.

Critical literacy[5] is a "central thinking skill" that involves the active questioning of ideas, moving beyond simply reporting on a text to analyzing and evaluating it. Rooted in the work of Brazilian educator

Paulo Freire, critical literacy seeks to analyze the relationship between language and power. Its purpose is to uncover embedded discrimination and challenge the power structures related to race, gender, and class that are often invisibly encoded in texts and media.

The New Literacy Studies (NLS)[6] framework challenges the "autonomous model" of literacy—the belief that literacy is a neutral, technical skill that automatically brings progress. In contrast, NLS proposes an "ideological model," which argues that literacy is always a social practice, embedded in specific cultural contexts and power relations.

When we apply these principles to technology, the implications are profound. Digital, media, and AI literacy cannot be seen as neutral skills. They are socio-technical systems "embedded with values, logics, and power structures." The narrative that AI will inherently improve education echoes the flawed "autonomous model." A critical approach, therefore, must move beyond "How do I use this tool?" to ask: "Who created this tool, and for what purpose?"; "Whose values are embedded in its design?"; and "Who benefits and who is harmed by its deployment?"

Locating AI Literacy: A New Tool or a New Way of Thinking?

The debate over how to teach AI literacy reflects the historical tension between functional skills and critical competency, with approaches falling along a spectrum.

At one end is the *instrumental or digital literacy model*, which frames AI literacy as the next logical step after digital and data literacy. This approach centers on proficiency with AI tools, aiming for effective and efficient use. Its primary goal is workforce readiness and task automation, with key skills including prompt engineering and using specific AI applications. From this perspective, AI is viewed as a neutral tool whose morality depends on the user, and the ethical focus is often narrowly defined by responsible use and avoiding plagiarism. The pedagogical goal is to teach students how to use AI.

At the other end of the spectrum, an emerging and more critical paradigm frames AI literacy as a fundamentally different kind of competency. This *critical and cultural literacy model* views AI not just as a tool, but as a complex socio-technical system that actively shapes culture, knowledge, and power. Its primary goal is to foster informed citizenship, ethical reasoning, and student agency. This approach recognizes the reciprocal relationship between AI and culture, where our values shape AI and AI, in turn, influences our cultural norms. It connects directly to social justice and critical pedagogy, empowering students to challenge how AI can reinforce racism, sexism, and other biases.

Consequently, the key skills are much broader, including bias detection, systems thinking, and epistemic judgment. AI is seen as an ideological artifact embedded with the values of its creators, and the ethical stance is systemic, addressing issues like data privacy, algorithmic bias, labor exploitation, and environmental impact. The pedagogical goal is not just to teach students how to use AI, but how and why to critique it, and—crucially—when not to use it.

This divergence shows that the debate over AI literacy is a proxy for a deeper conflict over the purpose of education. The instrumental approach aligns with a view of education as a pipeline for an AI-ready workforce. The critical and cultural approach aligns with a humanistic model focused on cultivating ethical and engaged citizens. The path we choose makes a powerful statement about our core educational philosophy.

The Heart of AI Literacy: Knowing When to Say No

If AI literacy is to be more than training on transient tools, it must be grounded in durable human capacities. The most significant challenges posed by AI are not technical; they are about our relationship with knowledge, our ability to think critically, and our capacity for sound judgment.

True literacy in any domain involves discretion. The pinnacle of AI literacy is the ability to make a conscious, critical decision about when *not* to use AI. This is not an act of technophobia but a reasoned choice. Reasons to forgo AI are numerous: concerns about data privacy, the risk of amplifying harmful biases, and the potential for generating "hallucinations." In our context as educators, the most compelling reason is pedagogical. We might choose to abstain from AI[7] to prioritize the development of foundational human skills: the "desirable difficulty" of brainstorming, the cognitive work of structuring an argument, or the personal process of finding one's authentic voice.

Generative AI operates on probabilistic pattern-matching, not factual understanding. This leads to profound challenges. It can produce "hallucinations," plausible-sounding but entirely false statements, with complete confidence. It is already being used to generate "synthetic data" in academic research, creating risks of misrepresentation. Some scholars argue this necessitates an "adaptive epistemology," where the primary intellectual skill is not knowledge acquisition but epistemic judgment: the ability to critically evaluate the credibility, context, and limitations of all information, especially that produced by an AI.

AI and Critical Thinking: A Double-Edged Sword

The relationship between AI and critical thinking is complex. On one hand, AI can be used to foster critical thinking by providing diverse perspectives or acting as a Socratic partner to challenge a student's reasoning. On the other hand, over-reliance on AI can lead to the atrophy of these same skills, encouraging a passive acceptance of machine-generated content. This is the fear that we are offloading the uniquely human task of thinking onto machines.

Navigating this requires cultivating intellectual virtues[8]: humility (recognizing AI's limits), courage (questioning AI's outputs), and curiosity (asking deeper questions). It is crucial to remember that AI itself cannot be virtuous; it is a non-conscious entity without moral agency. It can only exhibit "virtue-by-proxy," mimicking the ethical principles embedded by its human creators. This places the ultimate responsibility squarely back on our shoulders.

Teaching the Principles Without the Platform

If AI literacy is fundamentally about critical thinking, then its core principles can be taught effectively without ever using an AI tool. By focusing on the conceptual underpinnings of AI through "unplugged" activities, we can equip students with a durable skill set that transcends any specific technology.

"Unplugged[9]" activities are playful, often physical, learning experiences that teach computational concepts without computers.

Teaching Classification: In the "Good-Monkey-Bad-Monkey" game, students develop rules to classify images, creating a physical decision tree that teaches the logic of AI classification models.

Simulating Machine Learning: The "Monster Mapping" activity simulates a clustering algorithm by having students physically group monster cards based on shared features, learning how unsupervised machine learning identifies patterns.

Exploring Bias in LLMs: In "Large Language MadLibs," students use dice rolls to simulate how LLMs generate text based on word probabilities, demonstrating how biased data leads to biased outputs.

Engaging with Ethics: "Data Brokers" is a role-playing game where students act as companies using data to target users, sparking discussions about data monetization and surveillance.

These activities prove that the conceptual foundations of AI—algorithmic logic, the role of data, the emergence of bias—can be taught through tangible, interactive experiences.

Strengthening Foundational Literacies for the AI Era

The most effective preparation for a world with AI is to double down on foundational critical skills.

Advanced Source Evaluation: Traditional checklists like the C.R.A.A.P. test are insufficient for an AI-saturated landscape. We should instead teach more robust strategies like lateral reading, leaving a source to investigate its author and reputation elsewhere online, and the SIFT method (Stop, Investigate the source, Find better coverage, Trace claims to the original context).

Bias Detection in Media: Long-standing media literacy curricula provide a powerful foundation for understanding algorithmic bias. Lessons that teach students to analyze news sources for biased word choice, framing, and representation equip them with the lens needed to detect similar biases in AI-generated content. The core question of media literacy—"Who created this message and for what purpose?"—is precisely the core question of critical AI literacy.

Ethical Reasoning: Rather than being a niche topic, ethical reasoning should be integrated across the curriculum through project-based learning, scenario-based discussions, or by embedding ethics modules directly into technical courses.

When we do use AI tools, our pedagogy should center human agency, promote ruthless reflection on the tool's impact on our thinking, and use AI as a catalyst for deeper human inquiry.

AI Literacy as the New Humanism

A narrow, instrumentalist view of AI literacy—one focused only on using tools—is insufficient and potentially harmful. It risks producing compliant users of a technology they do not critically understand.

A more robust approach frames AI literacy as a critical and cultural practice. It recognizes that the core challenges of AI are not technical but humanistic: challenges of epistemology, ethics, and judgment. The most vital competencies are the abilities to critically evaluate information, discern truth from falsehood, understand how power is embedded in technology, and decide when *not* to delegate human cognition to a machine.

This reframing leads to a powerful conclusion: the most effective way to teach AI literacy is to strengthen the core of a humanistic education. The skills are not new. They are the timeless skills of critical thinking, close reading, and ethical reasoning. By teaching students how to analyze texts of all kinds (including algorithmic outputs) and question power structures (including those encoded in software), we equip them with a form of literacy that is truly future-proof.

Ultimately, the goal of AI literacy should not be to make students better at using AI, but to empower them to be more discerning thinkers, more ethical citizens, and more self-aware human beings in a world where AI exists. It is a call to reaffirm that the purpose of education is not to train operators for today's machines, but to cultivate the critical and creative minds needed to build a more just and thoughtful world tomorrow.

ORIGINALLY PUBLISHED ON THE AUGMENTED EDUCATOR
AUGUST 12, 2025

3

THE MOST ESSENTIAL SKILL IN THE AI ERA IS 2,500 YEARS OLD

A MODERN, YET SURPRISINGLY ANCIENT, FRAMEWORK FOR THE TIMELESS ART OF CRITICAL THINKING

I n a recent essay entitled "Beyond the Tool: Why True AI Literacy is About Critical Thinking, Not Prompting," I argued that the most productive way to approach AI in our classrooms is to frame AI literacy not as a technical skill, but as a critical and cultural practice. This perspective urges us to shift our focus from the fleeting mechanics of prompt engineering to the timeless, durable habits of mind we've always sought to cultivate: ethical reasoning, sound judgment, and above all, critical thinking.

But this argument immediately surfaces a deeper, more urgent question. We educators use the term "critical thinking" constantly. It's the North Star of our pedagogical mission, the glowing line item in every syllabus. But what do we actually mean by it?

This is not just an academic question. It's the central challenge of our time. We are caught in the great paradox of AI: a tool that can augment our intellect in profound ways, yet also tempts us into a state of "cognitive offloading[1]," where our own analytical muscles risk atrophy. Research is already showing some negative correlation between frequent AI use and critical thinking scores, as the friction-less convenience of AI encourages a mental passivity that is the very antithesis of the active, effortful inquiry we want for our students.

To navigate this, we need a definition of critical thinking that is robust enough for the digital age. And to find it, we need to look back. By tracing the 2,500-year evolution of this idea, we can see how humanity has repeatedly forged new intellectual tools in response to new challenges. This journey will show us that the answer isn't to invent a new form of thinking, but to reclaim and expand our most powerful traditions of inquiry for a world of new media.

Our guide for this exploration will be a model I've developed called the Four Lenses of Critical Engagement. This framework helps operationalize our rich definition of critical thinking for the modern world by extending the timeless principles of inquiry to the new forms of media that define our landscape: Critical Reading, Critical Listening, Critical Seeing, and Critical Making. As we trace the history of critical thought, we will see how these four lenses have always been present, and how understanding their evolution gives us a powerful map for the future.

The Ancient Art of Asking Why

Our journey begins, as it so often does, in the bustling agora of Athens with Socrates. His legacy is not a set of answers, but a method of relentless, disciplined questioning that revealed a foundational truth: a confident claim to knowledge is often a mask for an unexamined belief. The Socratic method is the origin of critical thinking as a practice designed to probe the very foundations of thought. Socrates taught us to be suspicious of confident, easy answers, a disposition more vital than ever when generative AI can produce them on any topic, in any style, in an instant. The Socratic spirit—the instinct to ask, "Why do you say that?" or "What are the consequences of that assumption?"—is the original, and still most powerful, antidote to the illusion of authority that AI projects.

This ancient practice[2] is the direct ancestor of two of the core competencies we need today. The Socratic dialogue, a shared exploration of ideas through speech, is the very essence of *Critical Listening*. It is the skill of not just hearing words, but of analyzing the underlying beliefs and value systems that shape them.

It was Socrates's student, Plato, who recorded these dialogues and, in doing so, drew a sharp distinction[3] between the Socratic pursuit of truth and the rhetoric of the Sophists, who used argumentation merely for persuasion. This ancient tension is startlingly modern. AI's capacity for "hallucinations," plausible sounding but entirely false statements, is a perfect digital reincarnation of

Sophistry: persuasive, but untethered from truth. Plato's insistence on truth over persuasion is the philosophical bedrock of *Critical Reading*, which demands that we move beyond a text's surface appeal to evaluate the validity of its claims.

Forging the Tools of Reason

If Socrates gave us the spirit of inquiry, the thinkers who followed sought to build its scaffolding. Aristotle, Plato's student, created the first systematic toolkit[4] for reasoning, the Organon, giving us the formal language of logic, deduction, and syllogism. He demonstrated that an argument has a structure, and that this structure can be analyzed and evaluated for its validity, separate from its emotional appeal. This was a monumental step, providing a universal grammar for assessing the soundness of an argument, whether we encounter it in an ancient text, a modern news report, or an AI-generated summary.

Centuries later, during the Scientific Revolution, Francis Bacon and René Descartes sparked a second revolution in thought. They argued that the source of authority for knowledge should not be ancient texts or abstract principles, but something new. For Bacon, it was empirical evidence, gathered through systematic observation and experimentation. His greatest contribution, however, may have been identifying the "Idols of the Mind," the innate biases that distort our perception of reality. This call for self-awareness is profoundly relevant today, as algorithms are often designed to exploit our cognitive biases, especially the confirmation bias that keeps us clicking on content we already agree with. Bacon's demand for empirical proof finds its modern echo in the practice of *Critical Seeing*: the skill of evaluating visual information, from a scientific chart to a viral video, and demanding that it corresponds to observable reality.

While Bacon looked outward, Descartes looked inward. He introduced "methodic doubt," a systematic skepticism[5] that refused to accept anything as true unless it could withstand rigorous questioning. This is the intellectual discipline we must now apply to the confident, authoritative voice of AI. The Cartesian method is the engine of modern critical inquiry, compelling us to deconstruct an argument, examine its premises, and test its logic before we grant it our belief.

Thinking as a Classroom Practice

For centuries, these were largely the tools of philosophers and scientists. It was the American educator John Dewey who carried them across the threshold of the modern schoolhouse at the turn of the 20th century. He reframed critical thinking, which he called "reflective thinking[6]," as the central purpose of education in a democracy.

Dewey defined it as the "active, persistent, and careful consideration of any belief... in the light of the grounds that support it." The word "active" is the key. It stands in direct opposition to the passive reception of information, a danger that has become exponentially greater in our age of endless scrolling and machine-generated content. Dewey's philosophy of "Learning by Doing"[7] was a call to engage students in the "desirable difficulty" of constructing their own knowledge through experience, not simply receiving it from an authority, whether that authority is a lecturing teacher or a large language model.

This Deweyan ideal is the direct intellectual foundation for the fourth and final lens of modern critical engagement: *Critical Making*. This is "Learning by Doing" for the digital age. It recognizes that creating content today requires a conscious, reflective engagement with our tools. It asks us to maintain human agency and ethical awareness, understanding how the tools we use shape our creative choices and how our creations participate in the wider information ecosystem.

A Framework for a New Era of Literacy

This journey reveals that critical thinking is not a single skill but a rich, composite practice. It is a Socratic disposition to question, an

Aristotelian toolkit for logic, a scientific commitment to evidence, and a Deweyan mandate for active, experiential learning.

The challenge posed by AI does not require us to invent a new form of thinking. It requires us to expand and adapt these foundational practices for a world where information is no longer confined to the printed page. This brings us back to the core argument of my AI literacy post: true literacy can be taught without ever touching the tool itself, because it is about cultivating a critical mindset. The Four Lenses Framework is a model for doing exactly that. It is not an "AI curriculum"; it is a comprehensive framework for critical thinking in a multi-modal world. AI is the catalyst that makes these skills urgent, but their scope is far broader.

Critical Reading: This is no longer just about analyzing a printed text. It's about interrogating the logic of hyperlinks, understanding the persuasive architecture of a website, and detecting the subtle biases in algorithmically curated news feeds. It's a foundational skill for navigating any information system, human or machine-made.

Critical Listening: This skill was essential long before AI-generated voices. It is the ability to analyze the rhetoric of a political speech, to detect the emotional manipulation in an advertisement's sound design, and to evaluate the credibility of a podcast host. In a world of synthetic audio, this auditory literacy becomes a crucial tool for verification, but its purpose is much larger: to understand how sound shapes belief.

Critical Seeing: We have always needed to be critical of what we see, from Renaissance propaganda paintings to 20th-century photojournalism. Today, this lens applies to deconstructing the narrative of a documentary film, interpreting the potential misrepresentations in a data visualization, and, yes, detecting the subtle impossibilities in an AI-generated image. It is the timeless skill of questioning all visual evidence.

Critical Making: Grounded in Dewey's philosophy, this lens applies to any act of creation. It encourages students to reflect on how a word processor's spell-check might homogenize their writing style, how a photo filter changes the emotional meaning of an image, or how collaborating with an AI can both open and close creative pathways. It is the practice of maintaining conscious, ethical control over the creative process, regardless of the tool.

By integrating these four modalities, we can move from a powerful definition of critical thinking to a tangible classroom practice. This approach doesn't ask us to abandon the core of a humanistic education; it asks us to strengthen it. It reaffirms that our goal is not to train students to be better users of AI, but to empower them to be more discerning thinkers, more ethical citizens, and more self-aware human beings in a world where AI exists.

That is a mission that Socrates, Bacon, and Dewey would surely recognize.

ORIGINALLY PUBLISHED ON THE AUGMENTED EDUCATOR
AUGUST 26, 2025

4

IT DOESN'T DO WHAT I WANT!

WHY TRADITIONAL COMPUTING EXPECTATIONS FAIL WITH ARTIFICIAL INTELLIGENCE

The more I engage with educators and professionals about AI integration, the more I encounter a telling complaint: "Whenever I try to use AI, it doesn't do what I want it to do." This frustration reveals something deeper than technical difficulty. It exposes a fundamental misunderstanding of how AI transforms our relationship with computing. Because rather than simply adding another tool to our digital toolkit, artificial intelligence introduces a radical paradigm shift that requires us to reconsider our basic assumptions about human-computer interaction.

Why Probabilistic Systems Behave Differently

The root of this misunderstanding lies in the contrast between two substantially different computing approaches. Traditional software operates primarily through deterministic algorithms: input a formula, receive a predictable result; apply a filter, see a consistent transformation. And while modern software increasingly incorporates probabilistic elements (spell-checkers suggest corrections, search engines rank results), these features are typically built around a fundamentally deterministic core.

By contrast, generative AI amplifies the probabilistic paradigm to an unprecedented degree. These systems don't execute deterministic processes, but instead generate outputs by analyzing statistical patterns derived from their training data. This means AI never guarantees an exact outcome; instead, it offers what it assumes to be the most probable response given the provided context. Understanding this distinction is crucial for anyone seeking to work effectively with these tools.

From Commands to Conversation

This probability-driven architecture transforms AI from a traditional tool into something resembling a collaborative partner. When we interact with AI, we initiate an exchange that unfolds through progressive refinement rather than single-shot execution. The system's first response can only ever serve as a draft, a starting point for dialogue rather than a finished product.

Consider a marketing designer creating a campaign banner. With conventional software, she manipulates pixels directly through precise commands. With AI, she might begin with a prompt describing her vision, receive an initial generation, then guide the system through several rounds of adjustment. She refines the color palette here, adjusts the composition there, perhaps regenerating specific elements while preserving others. After three or four iterations, she achieves a result that meets brand guidelines in half the usual time. The process resembles a conversation more than command execution.

This shift demands new skills: recognizing promising directions in initial outputs, identifying elements worth preserving, and articulating modifications that guide the system toward the intended outcome. Success depends less on memorizing commands and more on developing what we might call "conversational competence" with AI systems.

The Creative Professional's Adaptation

For those in creative fields, this transformation carries particular weight. Many designers, writers, and artists initially report feeling that AI threatens their expertise, their ability to translate vision into reality. Yet, this reaction often stems from approaching AI with deterministic expectations. When creators embrace the technology's collaborative nature, they discover it can accelerate exploration of creative possibilities, suggest unexpected directions, and enable rapid prototyping of ideas.

The key lies in reframing expertise. Rather than defining professional skill as the ability to execute a predetermined vision, we might understand it as the capacity to guide an iterative process toward innovative outcomes. This aligns with how creative work actually unfolds (through drafts, revisions, and experimentation) but now with an active artificial participant in that process.

Broader Implications for Digital Literacy

These lessons from creative workflows illuminate a broader transformation in professional computing. Just as designers must learn to guide AI through successive refinements, professionals across disciplines need to develop similar capabilities. This represents a significant shift in what we mean by digital literacy.

Traditional computer education emphasizes command mastery: learning specific features, memorizing shortcuts, following prescribed workflows. AI literacy requires different competencies: understanding statistical outputs, developing strategies for incremental improvement, and cultivating judgment about when an AI-generated result serves its purpose. Educational programs must evolve accordingly, teaching not just how to prompt AI systems but how to think probabilistically about their outputs and engage in productive iteration.

Additionally, modern AI tools offer some control over randomness through parameters like temperature settings and seed values, allowing users to balance creativity with consistency. Understanding these controls becomes part of the new literacy, helping professionals navigate between exploration and convergence as their projects demand.

Turning Frustration into Fluency

The complaint "AI doesn't do what I want" marks a transitional moment in our technological evolution. It reflects the discomfort of applying old mental models to new systems. As Donald Schön observed in *The Reflective Practitioner*[1], expertise often involves a conversation with the situation, and AI makes that conversation literal.

Those who thrive in this unfamiliar landscape will embrace the shift from deterministic to probabilistic thinking. They'll develop what we might call an iterative mindset, viewing each AI interaction as part of a design process rather than a failed command. They'll recognize that the technology's apparent limitation (its inability to read our minds and deliver perfect results immediately) actually opens new possibilities for discovery and innovation.

This transformation extends beyond individual practice to reshape entire fields. As more professionals develop fluency with probability-driven systems, we'll see alternative forms of human-AI and more general human-computer collaboration emerge. The question isn't whether AI can replace human creativity or judgment, but how we can orchestrate productive partnerships between human intention and machine capability.

By understanding AI's fundamental nature, not as a disappointing command executor but as a probabilistic collaborator, we position ourselves to harness its genuine potential. The future belongs to those who can navigate this new paradigm, guiding AI through thoughtful iteration toward outcomes that may surprise and exceed our initial visions. In embracing this shift, we take part in redefining what it means to work with computers in the twenty-first century.

ORIGINALLY PUBLISHED ON THE AUGMENTED EDUCATOR
JULY 5, 2025

5

THE EVOLVING ARCHITECTURE OF ARTIFICIAL INTELLIGENCE

UNDERSTANDING LLMS IN THE AGE OF LLAMA 4

The recent release[1] of Meta's Llama 4 has sparked renewed conversations about the underlying architectures that power large language models (LLMs). As educators increasingly integrate AI tools into their teaching practices, understanding these architectural differences becomes valuable—not just for technical knowledge, but for making informed decisions about which AI tools might best serve our educational needs. The shift toward

Mixture of Experts (MoE)[2] architecture in Llama 4 represents a significant evolution in how these models function, potentially offering new possibilities for classroom implementation while presenting unique considerations for educational contexts.

Unlike the opaque "black box" descriptions often used when discussing AI, the architectural foundations of these models reveal much about their capabilities, limitations, and potential applications in education. Whether you're seeking to understand why certain models perform better at reasoning tasks, why others excel at multilingual support, or how these systems might eventually run efficiently in resource-constrained educational environments, the architecture matters. Let's therefore explore the landscape of current LLM architectures and what they mean for educators navigating this rapidly evolving technology.

Transformer-Based Models

The transformer architecture[3] remains the fundamental building block of modern LLMs, serving as the backbone for virtually all models we interact with today. Introduced in 2017, this architecture revolutionized how AI processes language by allowing models to weigh the relevance of every word (or token) in relation to every other word—much like a teacher assessing each student's contribution during a class discussion. This mechanism, called "self-attention," was transformative—enabling models to capture long-range context and relationships in text in ways previous technologies couldn't.

When educators interact with models like OpenAI's GPT-4, Anthropic's Claude, Google's Gemini, or even open-source options like the Llama models, they're engaging with transformer-based systems. These models process text by encoding it into latent representations (essentially, turning words into numerical patterns that capture meaning) through stacked layers of so-called self-attention and feed-forward neural networks. This parallel processing approach allows the model to capture complex patterns in language while generating contextually relevant text.

Allrounder Models

For educators, these models function as effective all-around tools. Their extensive training on massive internet text datasets has equipped them with diverse skills and broad knowledge. This versatility allows them to be used in a wide range of educational contexts, such as explaining complex scientific ideas, creating creative writing prompts, translating languages, or helping computer science students with coding.

In multilingual classrooms, transformer models like Gemini and GPT-4 offer robust support across many languages, enabling educators to serve diverse student populations. And their interactive chat interfaces make them natural fits for tutoring scenarios, and smaller open-source transformer models can even run offline or on-premise —valuable for schools with data security concerns or unreliable internet access.

Transformer Limitations

However, these capabilities come with notable limitations. Dense transformer models (where "dense" refers to models where all parameters are active for every input) require substantial computational resources—the larger the model (more parameters), the greater the computing power needed. This creates a trade-off: more parameters yield better performance, but at significantly higher computational costs. For schools with limited budgets, this often means relying on cloud-based services with ongoing subscription fees rather than hosting models locally.

Additionally, while these models show impressive reasoning abilities, they still produce occasional "hallucinations" (incorrect information) and may generate biased outputs—requiring careful supervision from educators. And because their knowledge comes from their training data, they're unaware of current events unless specifically programmed to access external data.

The Architecture Behind Llama 4

Meta's recent release of Llama 4 represents a significant architectural shift toward what's known as Mixture-of-Experts (MoE). This approach tackles one of the fundamental challenges in scaling language models: how to increase model capacity without a proportional increase in computation.

How MoEs Work

The MoE architecture introduces multiple "expert" sub-models within a larger model framework, but it only activates a subset of these experts for any given input. In practical terms, an MoE architecture replaces certain layers of the transformer with "MoE layers" containing multiple parallel expert networks. A separate gating network (the "router") learns to choose which experts to activate for each input token. Most modern MoE implementations use what's called "sparse MoE," meaning only a small fraction of experts (perhaps 1 or 2 out of 16 total) are active for any given token.

This approach allows for dramatic increases in overall parameter count (and thus potential model capacity) without proportionally increasing the computation required for each input. For instance,

Llama 4 reportedly includes "16 experts" while effectively using only 17 billion parameters per input—suggesting a much larger total parameter count distributed across those experts.

Advantages for Educators

For educators, this architectural innovation offers several potential advantages. The promise of "more bang for the buck" in terms of model capability could translate to faster responses or the ability to run more capable models on limited hardware—important considerations for real-time classroom interactions or schools with constrained technology budgets.

The concept of specialized experts also aligns intuitively with educational needs. One expert might excel at step-by-step mathematical reasoning, another at coding assistance, and yet another at literary analysis. This could allow the model as a whole to cover diverse subjects more effectively than a traditional transformer model.

Challenges to Expect

However, MoE models also present unique challenges. Custom software frameworks are often needed to manage expert routing and parallel execution, adding complexity to model operation. Fine-tuning or prompting these models can be trickier; if the router isn't well-trained on your specific use case, it might not select the optimal experts. For educational settings where reliability and consistency are paramount, this added unpredictability requires consideration.

The memory demands can also be significant; all expert weights need loading regardless of simultaneous use. This could limit deployment options for educational institutions with older infrastructure, though techniques like expert parallelism and sharding can mitigate these challenges.

Chain-of-Thought Reasoning

As we explore LLM architectures, it's important to recognize that not all advancements come from novel network designs. Reasoning techniques like Chain-of-Thought (CoT)[4] have dramatically improved how these models perform on complex tasks without changing their underlying architecture.

Chain-of-Thought involves prompting or training the model to generate intermediate reasoning steps before providing a final answer—similar to asking students to "show their work" on mathematical problems. Rather than jumping directly to conclusions, the model articulates a step-by-step solution path, making its reasoning process explicit and transparent.

A Different Pedagogical Approach

For educators, this approach transforms how we can interact with AI in teaching contexts. When a model shows its reasoning process, both teachers and students can follow the logic and verify each step —crucial for subjects like mathematics, science, or critical thinking. It encourages methodical problem-solving and provides an opportunity to identify and address misconceptions or errors in reasoning.

Implementing Chain-of-Thought is remarkably straightforward: you can simply prompt a transformer model with phrases like "Let's think step by step" or provide examples of detailed reasoning before asking your question. Research has shown that this technique significantly improves model performance on complex tasks, particularly with larger models like GPT-4 or Google's Gemini.

The educational advantages are substantial. CoT makes AI responses more explainable, aligning with educational values of transparency and understanding. It transforms the AI from an answer provider into a model of thoughtful reasoning—demonstrating processes that students can learn from and emulate. It also reduces errors on complex problems by breaking them into manageable components.

Reasoning Models

Building on this concept, new reasoning-specialized models have emerged that incorporate CoT at their core. For example, OpenAI's ChatGPT o1 is trained to "think aloud" during problem-solving, showing each logical step. Anthropic's Claude 3.7 (featuring an "extended thinking" mode) similarly reveals its chain-of-thought for complex reasoning tasks. In the open-source space, DeepSeek R1—part of the DeepSeek V3 family—has been explicitly fine-tuned to produce multi-step CoT solutions by default. Each of these approaches places greater emphasis on transparent, step-by-step reasoning, making them particularly useful in educational contexts where process matters as much as the conclusive answer.

However, Chain-of-Thought makes responses longer, which might be tedious in some classroom scenarios. And even if the final answer is correct, the intermediate steps may be flawed, thus requiring teacher supervision when used as examples. Finally, while the technique shows dramatic improvements with larger or specialized models (like ChatGPT o1, Claude 3.7, or DeepSeek R1), smaller models may lack the capacity to carry out multi-step reasoning coherently.

Tool Use in LLMs

Even the most sophisticated language models face inherent limitations: they possess only the knowledge contained in their training data, they can't independently verify facts, and they may struggle with complex calculations or specialized tasks. To address these constraints, researchers and companies have developed methods for LLMs to use external tools, effectively allowing them to augment their capabilities beyond what's encoded in their parameters.

A tool-using LLM can recognize when it needs additional information or specialized processing and can then invoke appropriate external systems such as search engines, databases, calculators, or code interpreters. This dramatically expands what the model can do and improves reliability for certain queries.

Educational Applications

In educational settings, tool use essentially transforms an LLM into a multi-functional assistant. A model with search capabilities can provide up-to-date information about recent scientific discoveries or current events, ensuring that classroom discussions remain relevant and accurate. Models with calculation tools can solve complex mathematical problems with precision, while those with code execution abilities can show programming concepts through working examples.

This approach addresses several key concerns for educational applications. It ensures knowledge remains current beyond the model's training cutoff date, which is critical for rapidly evolving fields. It improves accuracy and verification, allowing models to check facts rather than relying solely on internal knowledge. And it enables more interactive learning experiences through integrating specialized tools for tasks like image generation, graphing capabilities, or data visualization.

Added Complexities

However, tool use introduces its own complexities. It typically requires internet access or APIs to external services, which may involve costs or technical setup. There are security and privacy considerations, particularly when allowing models to search the web or execute code in educational environments. And reliance on external tools means the overall system is only as reliable as those tools and the internet connection supporting them.

For educators considering tool-augmented LLMs, these trade-offs require thoughtful consideration. The advantages of the improved capabilities should be carefully considered alongside any potential risks, with proper safeguards and supervision implemented for safe and productive classroom use.

Current Models and Their Architectural Approaches

Understanding the landscape of current models and their architectural choices helps educators make informed decisions about which AI tools might best serve their needs. Below are some prominent examples[5]:

GPT-4 (OpenAI): While OpenAI hasn't confirmed specific architectural details, industry analysis suggests GPT-4 may employ a Mixture-of-Experts approach internally. This would help explain its impressive reasoning capabilities across diverse domains. GPT-4 supports tool use through plugins and shows strong Chain-of-Thought reasoning when prompted. For educators, GPT-4 offers comprehensive capabilities, though access involves subscription costs.

ChatGPT o1 (OpenAI): A new reasoning-focused model from OpenAI's "o-series," ChatGPT o1 prioritizes methodical, step-by-step problem-solving, showing intermediate reasoning by default. It has proven especially adept at math and coding tasks. Although its coverage of general knowledge is narrower than GPT-4, it excels in transparent explanation, which is highly valuable in classrooms, emphasizing the process over mere answers.

Llama 4 (Meta): Meta's newest model family embraces the MoE architecture explicitly, reportedly using "16 experts" while effectively using 17 billion parameters per input. This approach enables performance gains without proportional increases in computation. Since it has been released openly, it could offer educators who have access to the required computational resources more accessible options for deploying advanced AI locally.

Claude 3.7 (Anthropic): Anthropic's latest version of Claude builds on a dense transformer core but emphasizes aligned, transparent reasoning. It features an "extended thinking" mode where the model reveals its chain-of-thought for complex tasks, helping students see how it works through solutions step by step. The large context window also remains a signature feature, enabling the analysis of lengthy documents in one go.

DeepSeek V3 & R1 (Open Model): An open-source project showcasing a powerful Mixture-of-Experts design (V3) and a specialized chain-of-thought variant (R1). DeepSeek R1 is explicitly fine-tuned for multi-step reasoning, producing detailed solution paths by default. This openness, coupled with strong performance in math and coding tasks, makes it attractive for educational institutions seeking cost-effective and customizable AI solutions.

Gemini (Google): Following PaLM 2, Google introduced Gemini as its flagship multimodal transformer. It extends beyond text to include image and audio processing and is built to handle extremely large context windows. Gemini continues Google's strong multilingual tradition and offers built-in tool use, making it a capable option for diverse or rapidly changing classroom scenarios.

Mixtral 8×7B (Mistral AI): An example of an open MoE model. Despite a modest overall size compared to larger closed-source models, its performance has surprised researchers. For educators wishing to run AI locally, Mixtral demonstrates the potential for combining experts for efficiency and specialization.

Choosing the Right Architecture for Your Needs

As we navigate this complex landscape of model architectures, the key is aligning technology choice with educational objectives and practical constraints. Below are some considerations to help match various architectures and features to real-world classroom scenarios:

General Classroom Support: Dense transformer models like earlier Llama versions or Gemini are excellent all-purpose solutions when you need broad knowledge across many subjects. Their ease of use and established integration pathways can be invaluable if you're just starting with AI in your curriculum.

Specialized Subject Teaching: MoE-based models—exemplified by Llama 4 or DeepSeek V3—can potentially leverage domain-specific experts for tasks like math, coding, or language arts. If your teaching involves diverse topics and you want a single AI capable of specialized assistance, these architectures may be able to deliver more focused performance without requiring massive computational resources.

Transparent Reasoning and Critical Thinking: If encouraging students to observe and engage with detailed thought processes is a priority, consider a reasoning-optimized model such as ChatGPT o1, Claude 3.7, or DeepSeek R1. Their chain-of-thought focus provides step-by-step clarity, helping learners identify logical approaches and potential pitfalls—valuable for subjects like science labs or advanced mathematics.

Real-Time Research Projects: Tool-augmented models, whether GPT-4 with plugins or open equivalents, can confirm facts, generate up-to-date references, or handle data-driven tasks on the fly. This capability is useful for project-based learning where current events or live data are central, although it introduces extra technical steps for setup.

Resource-Constrained Environments: While powerful dense models still demand significant computational resources, the efficiency gains promised by MoE approaches might eventually allow advanced AI to run locally on more modest hardware. Alternatively, smaller open-source transformers can be deployed on-site if access to commercial APIs is infeasible, ensuring data privacy and cost control.

Balancing Trade-Offs: Each architecture comes with a unique set of benefits and drawbacks—MoE models may require more intricate prompting, while dense transformers can be simpler but less efficient at scale. Reasoning-centric models produce thorough step-by-step explanations but at the cost of longer processing times. Tool use offers richer capabilities but relies on stable internet and external integrations.

Ultimately, no single model or architecture is universally "best." The decision hinges on your classroom's specific needs: the nature of the subject, the depth of reasoning you want to emphasize, the level of privacy or internet access you can afford, and the funds or infrastructure at your disposal. Thoroughly assessing these factors ensures you can deploy the most suitable AI assistant for your learning community.

The Journey Ahead

Meta's release of Llama 4 with its Mixture-of-Experts architecture signals an important evolution in how language models are designed and deployed. Rather than simply scaling up existing approaches, researchers are exploring architectures that balance capability with efficiency, potentially making advanced AI more accessible for educational contexts with limited resources.

Looking ahead, we can expect further architectural refinements that address current limitations. Models may become more modular, allowing educators to select specific capabilities based on their needs. Reasoning techniques will likely continue evolving, enabling AI to tackle increasingly complex problems with greater transparency. And tool integration will expand, connecting models to specialized educational resources and platforms.

For now, understanding the landscape of current architectures—from transformers to MoE, from chain-of-thought reasoning to tool use—empowers us to select AI tools that align with our pedagogical goals. By remaining curious about these developments while keeping our focus on student outcomes, we can navigate this rapidly evolving technology landscape as truly augmented educators.

ORIGINALLY PUBLISHED ON THE AUGMENTED EDUCATOR APRIL 13, 2025

6

THE EVOLUTION OF EDUCATIONAL VALUES

A PERSONAL PERSPECTIVE ON AI AND WRITING

R ecently, I've observed an increasing number of educators sharing their thoughts on SubStack about integrating AI into teaching creative writing. A common thread emerges: the concern that AI technologies might undermine traditional educational values and skills. This anxiety reflects a broader discussion about the changing nature of education and the role of human creativity in an increasingly digital world.

Matthew Kirschenbaum's concept[1] of the "Textpocalypse," introduced in an article in The Atlantic, captures these concerns effectively. He describes an impending "tsunami of text and content" generated by AI systems, raising valid questions about the future value of human-authored writing and the increasingly complex relationship between human and machine-generated content.

I deeply empathize with educators who have dedicated their careers to fostering human authorship and creative writing and now fear the loss of many of these values. However, drawing from my personal experience with radical technological change, I'd like to offer a different, and potentially controversial, perspective on this issue.

What if our current valuation of human-authored writing merely represents a transient phase in our intellectual evolution?

The skills and values we consider essential in education have never remained static. Many capabilities once deemed crucial decades ago now hold little practical relevance. This pattern of change suggests that our present educational priorities even within the context of something as fundamental as creative writing may not maintain their significance indefinitely.

A Personal Journey Through Educational Change

My career path illustrates how dramatically educational values can shift over time. I began in a field that's barely recognizable today, as technological advances have fundamentally transformed the skills I acquired during my graduate studies and early academic work.

In the mid to late 1980s, I studied at the Vienna University of Technology, specializing in the pedagogy of "Descriptive Geometry" - a discipline with a rich tradition also known as the "Vienna School of Geometry." Our role was to teach engineering and architecture students the precise art of hand-drawn technical drawings.

It is important to point out that the curriculum demanded extraordinary precision and patience. For example, engineering students learned to determine the exact radius of curvature in cylindrical intersections using only ruler and compass. And architecture students mastered every aspect of three-point perspective techniques for accurate architectural renderings of complex curved surfaces.

This was a time-consuming craft. A single drawing might require days of careful and laborious work, culminating in a final ink rendering where a single mistake could cause starting over completely. Unfortunately, I no longer have any of my own drawings. The University archive kept the better ones, and I lost the ones I had during a move long ago.

What I do have, however, is a video recording of one lecture I gave as a student in 1989. You can see some screenshots included here. And if you don't mind the inferior quality recording and German language, you can watch my full lecture on YouTube[2], where I published it as an unlisted video.

I also still have the compass I used in this video; you can see it in this post's cover image. The compass was custom made by a local carpenter so that it would provide the required accuracy on the blackboard. And in case you are wondering, we wore lab coats during our lectures. This wasn't for show - we used colored chalk on the blackboard, and these stains were impossible to remove from street clothes. The lab coats were essential for protecting our clothing.

The Parallel with Today's AI Transition

As computer-aided design (CAD) systems became more accessible in the mid-1980s, our initial response mirrored current reactions to AI in writing. We argued that hand-drawing skills were fundamental to developing critical evaluation abilities in engineers. We believed architects couldn't truly understand spatial relationships without manual drawing experience. These arguments sound strikingly similar to current debates about AI and writing.

However, technological progress gradually and relentlessly challenged these assumptions. Our department eventually merged with mathematics, and our courses were significantly reduced in scope. I adapted to these changes, moving on to new challenges that bore little resemblance to my original training.

Today, courses about descriptive geometry remain in the curriculum in many universities, but they are now specialized components of engineering and architectural education, not fundamental ones. And I sometimes wonder if these courses focus more on preserving cultural heritage than on teaching practical skills.

Lessons for the Present

Nevertheless, the quality of engineering and architectural work hasn't diminished with the transition to CAD systems. Instead, these tools have enabled new possibilities while maintaining professional standards. And our educational values have evolved alongside these technological changes.

Engineers can now model complex mechanical systems with unprecedented precision, testing multiple iterations before production. Architects design buildings that would have been nearly impossible to conceptualize with traditional drawing methods. The tools have changed, but the fundamental understanding of spatial relationships and engineering principles remains crucial - it's just expressed through different means.

I believe this history can offer a valuable perspective on current debates about AI and writing. Just as the importance of hand-drawn technical drawings proved less critical than we imagined, perhaps we should examine our assumptions about the absolute necessity of traditional human authorship in writing.

The evolution of educational values isn't about loss, but transformation. While certain traditional skills may become less central, new capabilities and forms of creativity emerge. The key lies in maintaining an open mind while ensuring we prepare students for the future they'll actually encounter, not the past we're familiar with.

This perspective doesn't aim to dismiss current concerns about AI in education, but to suggest that a broader historical lens could benefit our view of these changes. Perhaps the question isn't whether AI will devalue human writing, but how it might transform our understanding of authorship and creativity in ways we haven't yet imagined.

ORIGINALLY PUBLISHED ON THE AUGMENTED EDUCATOR DECEMBER 27, 2024

7

THE PROFESSIONAL'S PARADOX: WHY CREATIVE INDUSTRY EXPERTS GET AI DISRUPTION WRONG

THOSE WHO JUDGE TODAY'S AI BY YESTERDAY'S STANDARDS RISK MISSING TOMORROW'S REVOLUTION

I recently came across a comment by an established visual effects professional, who dismissed a video made with a new generative AI tool as "utter bull" because it had a subtle slow motion feel to it. Another industry expert called the technology "absolutely dog shit—an insult to creative and artistic fields" for similar reasons. Such arguments are becoming a familiar refrain in professional circles: the motion is inconsistent, the physics are wrong, it lacks fine-grained control, and the results feel soulless, uncanny, or just plain weird. In short, it's not up to professional standards.

From a psychological perspective, this reaction is completely understandable. When a new technology emerges that threatens to automate skills honed over a lifetime, it's natural to feel defensive. Livelihoods are on the line, and there is a genuine fear that the nuance, dedication, and hard-won expertise of human artistry are being devalued. This isn't just about jobs; it's about identity.

However, while the sentiment is understandable, the argument itself is dangerously flawed. It reveals a fundamental misunderstanding of how transformative technologies actually work.

Interestingly, the main reason creative industry experts so often get this wrong isn't due to a lack of intelligence or foresight. It's because the very principles of professional expertise that made them successful in the first place become strategic blind spots in the face of disruption. Their finely-tuned instincts, trained to maximize professional quality, actively steer them away from the messy, low-margin, and initially inferior world where disruption is born.

The "Good Enough" Revolution

The late Harvard professor Clayton Christensen[1] called this phenomenon the "innovator's dilemma." His theory of disruptive innovation doesn't describe new technologies that are immediately better than existing ones. In fact, it's the opposite.

Disruptive innovations almost always start out as inferior when measured by the performance metrics that matter to the most demanding customers. They don't win by stealing the incumbent's best clients. Instead, they take root by creating new markets of people who were previously non-consumers or by serving "overserved" customers at the low end of the market with a simpler, cheaper product. For this initial audience, the new technology is, plain and simply, "good enough."

This is precisely why established experts misjudge it. Their entire business model is built on delivering high-quality products to their most demanding customers. When faced with a low-quality, low-margin alternative, the rational decision is to ignore it and often "flee upmarket" toward even higher requirements. The processes designed to kill bad ideas also effectively kill disruptive ones.

As an example, think of the shift from analog film to digital photography. Early digital cameras[2] were, by every professional metric, terrible. They offered laughably low resolution, poor color fidelity with "hideous magenta color casts," sluggish autofocus, and terrible battery life. Professionals, who had mastered the art of film, belittled the new technology. They argued digital could never replicate the "film look" and that it "reduced the skill requirement."

They were judging the new technology against the standards of the old. But early adopters of digital weren't using it to do the same "job" as film. They were using it for its new, disruptive advantages: the immediacy of an LCD screen, the convenience of no darkroom or chemicals, and the zero marginal cost of taking another picture. It was good enough for a massive new market of casual users. Kodak, by listening to its high end customers who wanted better film, famously missed the disruption[3] entirely. It wasn't a failure of management; it was a failure born from the logic of good management.

What "Job" Is Generative AI Being Used For?

This brings us back to generative AI video. The critiques about its lack of professional quality are the modern-day echoes of photographers dismissing digital's low resolution or typesetters decrying the "ransom note effect" of early desktop publishing. They are all symptoms of judging a disruptive technology by sustaining standards.

The critical question is not, "Is generative AI as good as a team of VFX artists today?" The question is, "What job is it being used for that was previously impossible or too expensive?" When we look closely, we see generative AI finding its footing not by replacing high-end production, but by tackling jobs that were previously underserved or didn't exist at all. This includes:

Rapid Ideation and Pre-visualization: Before a single frame is shot, filmmakers, ad agencies, and game developers spend enormous resources on conceptualization. Generative AI is being used for the job of "help me see my idea, right now, for a fraction of the cost." This isn't about creating the final, polished product; it's about rapid, low-cost prototyping that unlocks more creative exploration upfront. The technology serves as a tireless creative partner, helping artists overcome creative blocks by providing novel starting points and unexpected visual combinations.

Creating Low-End and Commodity Content: There is a vast universe of video content that doesn't require a blockbuster budget. Think of social media ads, internal corporate training videos, or SEO-optimized blog content. For these applications, traditional production is often too slow and expensive. Generative AI is a classic low-end disruptor, providing a tool that is perfectly good enough for this high-volume, low-cost segment. This is a market that high-end VFX studios have little financial incentive to fight for, allowing the disruptive technology to quietly capture a significant foothold.

Democratizing Creation: Perhaps its most powerful disruptive function is enabling what Christensen called "non-consumption." There are millions of people who have stories to tell but lack the budget, equipment, or technical skill for traditional video production. Generative AI is being used by students, indie musicians, teachers, and small business owners who previously had no access to motion graphics or visual effects. The AI is competing with the alternative of having no video at all, creating a vast new market of creators and fundamentally leveling the creative playing field.

These are the footholds. From here, the technology's S-curve of improvement is incredibly steep. Models like Sora, Veo, Kling, and Runway are advancing at a breakneck pace, and what seems flawed today will be good enough for more demanding tasks tomorrow[4].

The Danger of Dismissal

To judge today's generative AI by the standards of a mature, professional creative pipeline is to fundamentally miss the point of the disruption that is occurring. It's focusing on the technology's current weaknesses while ignoring its unique, game-changing strengths: speed, accessibility, and radically lower cost.

There is real danger is dismissing the technology because it's not yet perfect. Professionals and educators who cling to the belief that "it's not good enough" are falling into the classic trap of the innovator's dilemma. The very paradigms of skilled expertise that drive success become counterproductive when facing a disruptive technology. They create blind spots, making it rational to ignore the "toy" that is quietly but surely redefining the market from the bottom up.

The dismissal is a path to obsolescence and the mindset therefore has to shift from viewing AI as a threat that takes something away, to seeing it as a tool that adds to what you do. The choice isn't between maintaining the status quo and adopting AI; it's between adapting or being left behind.

But adaptation also doesn't mean abandoning core creative skills. Rather, it means augmenting them by developing expertise in AI tool selection, learning the art of AI direction, and focusing on the uniquely human abilities of strategy, taste, and emotional storytelling that machines cannot replicate. The artists who thrive will be those who learn to collaborate with these new tools, not those who pretend they don't exist.

Instead of asking whether it meets our current standards, we should be asking what its disruptive capabilities tell us about the future of the creative industries. The value is shifting from pure technical execution—the "how"—to creative direction, taste, and storytelling—the "what" and the "why." The most valuable creative professionals of the future won't be the ones who can outperform the machine at technical tasks. They will be the ones who know how to direct it, acting as a creative collaborator with a powerful new tool.

As educators, our role is to prepare our students for this augmented future. We must teach them not just the craft of today, but the strategic foresight to see where the tools are heading tomorrow.

Because the disruption isn't coming. It's already here.

ORIGINALLY PUBLISHED ON THE AUGMENTED EDUCATOR AUGUST 19, 2025

PART II

NOTES FROM THE FRONT LINES

ON TEACHING IN THE AGE OF AI

With the foundational concepts established, we now pivot from the theoretical to the tactical. This section moves the conversation into the classroom, where the abstract questions of the AI revolution collide with the messy realities of teaching, curriculum, and institutional practice.

The essays that follow are dispatches from these front lines. They confront the immediate challenges facing educators today: the emergence of a new "AI productivity divide" that threatens to deepen inequity, and the collapse of traditional academic integrity in a "post-plagiarism era." Yet, this is not a collection of warnings, but a record of active engagement. Here you will find personal accounts of using AI in grading and creative production, alongside field-tested frameworks designed to make the thinking process, not the final product, the center of our assessments.

These are blueprints and hard-won insights for fellow educators in the midst of this transformation.

8

THE HIDDEN INEQUITIES OF AI IN EDUCATION

A NEW DIGITAL DIVIDE: WHO'S REALLY GETTING AHEAD?

I recently came across Jason Hamilton's fascinating YouTube channel[1], "The Nerdy Novelist,"where he teaches writers how to use AI tools to enhance their craft. His journey exemplifies the transformative potential of AI. After traditionally writing over a dozen fantasy novels—a process that took months of meticulous crafting, rewriting, and polishing—he began experimenting with AI tools. The results were astonishing: he is now able to complete full length novels in days instead of months, all while maintaining the quality his readers expect.

But this post isn't about Jason or novel writing. What strikes me is that his experience captures something I see rapidly emerging in education as well: the "AI productivity divide." We're starting to witness a profound split between those who can effectively leverage AI tools and those who cannot—a divide that could redefine educational equity in unprecedented ways.

Beyond Hardware: The New Digital Frontier

The days when access to laptops defined technological equality in education feel like ancient history. Today's divide runs deeper—it's about mastering AI as a cognitive tool. Some students are becoming fluent in working with AI, developing an intuitive understanding of how to leverage these tools for learning and problem-solving. Others are left behind.

This technological gap represents something fundamentally different from previous digital divides. While earlier advances in educational technology offered steady but incremental improvements, AI tools can exponentially expand a student's capabilities. Students who effectively use AI aren't just working faster—they're working differently, engaging with material more deeply, and developing more sophisticated understanding in less time. This efficiency creates a compounding advantage as they use their extra time to explore advanced concepts or develop new skills.

The Multiplier Effect: How AI Is Reshaping Learning

The AI divide transforms learning at every level. To understand its profound impact, let's follow two graduate students preparing literature reviews for their thesis proposals:

Doris, who is AI-literate, begins by engaging in a sophisticated dialogue with her AI assistant. She starts by exploring how key theories in her field have evolved over the past decade. The AI helps her identify major theoretical shifts and emerging debates, suggesting connections between different schools of thought. When she encounters unfamiliar methodologies or conflicting interpretations, she prompts the AI to explain the underlying assumptions and potential implications. The AI also helps her recognize patterns across numerous studies, highlighting where consensus exists and where significant questions remain. Within hours, she has developed a clear understanding of her field's theoretical landscape and identified promising directions for her own research.

Meanwhile, Michael approaches the same task traditionally. He begins with recent publications, following citations backward through the literature. When he encounters competing interpretations or methodological debates, he must pause to dive into multiple papers, often losing sight of his original research question. He spends days navigating academic databases, trying to determine which papers represent fundamental contributions and which are minor variations on established themes. Without interactive guidance, he struggles to see how different theoretical approaches relate to each other. His notes are comprehensive but disconnected, making it difficult to construct a coherent narrative of how his field has developed.

But research skills are just the beginning. AI-literate students approach problem-solving in revolutionary ways. They explore multiple solutions simultaneously using AI while receiving immediate feedback to adjust their learning strategy. Through AI-enhanced iteration, they develop increasingly sophisticated work. As they progress, they discover unexpected connections across different subjects, building a more integrated understanding of their studies.

The technology isn't just making them faster. It's fundamentally changing how they think, learn, and create.

Bridging the Divide: A Call to Action

This divide could create one of the most significant educational inequalities in generations. However, there's a path forward provided that we act now. Here's what I think needs to happen:

First, we need universal AI literacy programs integrated into every curriculum. This isn't about teaching specific tools but cultivating an "AI mindset" that empowers students to leverage AI effectively. Just as we teach critical thinking and research skills, AI literacy should be a fundamental component of education.

Second, we must rethink how we assess learning. Traditional assessments often miss the depth of learning and creative problem-solving AI enables. We need new frameworks that encourage students to use AI as a tool for deeper understanding rather than a shortcut to traditional metrics.

Finally, we need robust support systems to identify and help students who struggle with AI integration. This means developing early warning systems, providing targeted interventions, and ensuring every student has access and guidance to thrive in an AI-enhanced learning environment.

The Immediate Horizon

The AI productivity divide is already reshaping how humans learn and create. In the immediate future, we'll see this impact ripple through our educational institutions. Students who master AI integration will quickly gain advantages in problem-solving, creativity, and productivity. Their capacity for innovation will be dramatically enhanced, positioning them to tackle increasingly complex challenges.

Without intervention, these advantages will compound rapidly. Educational institutions that successfully integrate AI literacy programs will see their students pull ahead, while those that delay may find their students struggling to catch up. The gap between AI-literate and AI-naive students will widen with each passing semester.

The Long-Term Stakes

Looking further ahead, the implications become even more profound. This divide will reshape career opportunities, social mobility, and economic equality for decades to come. Students who master AI integration will enter the workforce not just with technical skills, but with fundamentally different approaches to problem-solving and creativity.

The social impact could be transformative—or devastating. Without intentional intervention, the AI divide threatens to calcify existing social inequalities and create new barriers to upward mobility. Yet communities that successfully bridge this divide will likely see accelerated economic development and increased innovation, creating new opportunities for growth and prosperity.

The path we choose now will determine whether AI becomes a great equalizer in education or another barrier to opportunity. We need a coordinated effort from educators, policymakers, and technology leaders to ensure AI literacy becomes as fundamental as reading and writing.

The future of education isn't just about having AI in classrooms— it's about ensuring every student can harness its power effectively. By acting decisively now, we can shape an educational landscape where AI enhances learning for all students, not just a privileged few.

We're at a crossroads. We can let AI deepen existing educational inequalities, or we can use this moment to create a more equitable and effective educational system for all students. The choice—and the responsibility—is ours.

ORIGINALLY PUBLISHED ON THE AUGMENTED EDUCATOR
OCTOBER 30, 2024

9

THE EMERGING ACHIEVEMENT GAP IN EDUCATION

AI LESSONS FROM MY GAME DESIGN CLASSROOM

L ast week, while watching a YouTube video by Jason Hamilton[1], a.k.a. "The Nerdy Novelist," about AI's role in writing, something clicked. In the video Jason emphasized the importance of "working harder *and* smarter" with AI, not just working smarter. This simple yet profound insight perfectly captured what I've been observing in my graduate-level game design classroom, where the impact of AI tools has created the largest achievement gap I've ever seen in my teaching career.

When AI Meets the Classroom: A Teaching Experiment

Last term, I had decided to run an experiment in my introductory game design course for graduate Digital Media students. Instead of discouraging the use of or even banning AI tools – as many educators have chosen to do – I actively encouraged their use. To assist the students, I created video tutorials showing how to integrate AI throughout the game development process.

Here's a practical example from my class: In previous terms, creating a playable 3D game character would take my students several weeks. The process involved character design, 3D modeling, rigging (creating a skeleton for animation), and implementing animations. This term, students could use a chain of AI tools – MidJourney for character design, 3DAIStudio for 3D modeling, Adobe Mixamo for rigging and animation, and Spline for web implementation – to complete the same task in under an hour.

The results? They weren't quite as polished as traditionally crafted characters, but they were more than serviceable for learning game design principles. More importantly, this efficiency allowed students to focus on other aspects of game design they might never have had time to explore.

The Tale of Four Students

To understand what happened next, let me introduce you to four types of students I observed in my class. Similar to how Jason Hamilton described the phenomenon in his video, I am representing them with four fictional characters:

1. Larry - Puts in minimal effort and refuses to use AI tools
2. Angie - Puts in minimal effort but embraces AI tools
3. Harry - Works hard but refuses to use AI tools

4. Sarah - Works hard and embraces AI tools

It is obvious that different students will perform differently in every class. However, in previous terms, the quality gap between my highest and lowest performing students was not all that dramatic. This term? The difference was staggering. It was like comparing middle school projects to professional game development work. What I witnessed in my classroom this term was an achievement gap I had not seen before.

The Unprecedented Achievement Gap

Many educators are primarily concerned about students like Angie who use AI with minimal effort to match the output of traditionally hardworking students like Harry. But in doing so they're missing something far more significant: the enormous divide that's emerging between students who combine minimal effort with AI resistance or unfamiliarity and those students who embrace both hard work as well as new technologies.

The contrast between Larry and Sarah's work was nothing short of stunning. Larry, putting in minimal effort and refusing to engage with AI tools, produced work that barely met basic requirements. Sarah, on the other hand, combined her strong work ethic with AI's capabilities to create projects that wouldn't look out of place in a professional portfolio. She used the time saved by AI tools to push boundaries, experiment with advanced concepts, and refine her work to a level I've never seen before in an introductory course.

This gap isn't just larger than usual – it's fundamentally different in nature. While educators worry about AI tools letting less motivated students take shortcuts, we're overlooking how these same tools, when combined with dedication and hard work, are enabling our most committed students to soar to new heights. The distance between Larry and Sarah's work isn't just a matter of degree; it's a complete paradigm shift in what students can achieve.

Why This Matters Beyond the Classroom

It is important to note that this isn't just about grades or classroom performance. We're witnessing the emergence of a new kind of performance divide – one that's not based on traditional measures of academic ability or effort alone, but on students' willingness and ability to effectively leverage AI tools.

And let's not forget that in the professional world, the ability to work with AI isn't just an advantage anymore – it's becoming a necessity. By preventing students from learning how to use AI tools in educational settings, we might think we're maintaining equity and fairness, but we're actually doing them a disservice. We're denying them the opportunity to develop crucial productivity skills they'll need in their careers.

The New Educational Challenge

I strongly believe that as educators, we need to shift our focus. Instead of debating whether to allow AI tools in our classrooms, we should be asking ourselves:

- How do we help students like Larry not only embrace new technologies, but also discover how these tools can motivate them to work harder?
- How do we ensure students like Angie develop and maintain strong fundamental skills while leveraging AI?
- How do we show hardworking students like Harry that AI tools can amplify their efforts rather than replace them?
- And perhaps most importantly, how do we create learning environments that encourage more students to become like Sarah?

The achievement gap I witnessed in my game design class isn't just a warning sign – it's a glimpse into the future of education. Our challenge isn't to prevent students from using AI tools; it's to ensure all students can use these tools effectively while maintaining their drive to excel.

The future belongs to those who can work both harder and smarter. As educators, it's our responsibility to prepare all our students for that future, not just the ones who figure it out on their own.

ORIGINALLY PUBLISHED ON THE AUGMENTED EDUCATOR
NOVEMBER 2, 2024

10

RESISTANCE IS FUTILE

ACADEMIC INTEGRITY IN THE POST-PLAGIARISM ERA

This year I attended the INTED2025 conference in Valencia, a global gathering which offered a revealing window into how quickly educators worldwide are adapting to AI-driven technological change. One ubiquitous theme at the conference demands further examination: the future of academic integrity. We now enter what scholars have aptly termed a "post-plagiarism era"— one where our traditional understanding of academic dishonesty requires fundamental reconsideration. And very few educators are ready for this.

The concept of "post-plagiarism" emerged relatively recently in academic discourse through Sarah Elaine Eaton's 2021 book *Plagiarism in Higher Education*[1]. This framing acknowledges a fundamental shift: our traditional understanding of plagiarism—defined primarily as verbatim copying without attribution—has become inadequate for our new reality. With AI now capable of creating seemingly original content without direct copying, academic integrity faces a fundamental challenge in definition, detection, and response.

In this emerging landscape, human-AI co-authorship becomes the norm rather than the exception. The argument is that our focus must therefore shift away from simply running students' work through plagiarism checkers toward understanding and managing the responsible use of AI tools. Most importantly, we need to expand the conversation about plagiarism from its narrow definitions to embrace the broader concepts of ethical learning and academic honesty.

An Experiment in Futility

To explore the underlying key issue of this debate, I conducted a simple experiment testing the effectiveness of conventional plagiarism detection systems with AI-assisted writing. To be clear, this isn't part of any formal research—I simply wanted to determine whether our plagiarism tools could still detect AI-generated academic dishonesty and how easily these systems might be circumvented.

Hint: it turned out to be exceptionally easy.

I started by selecting the following paragraph from one of my most cited papers—a 2005 work[2] on the scientific relevance of eSports:

> It has to be noted that there is no particular need to look at eSports
> as an area of disciplines that satisfy a traditional sport definition. We
> could just as well look at eSports as a completely separated field of

study. The overrated question whether competitive gaming is a sport or not is to some extent irrelevant for the academic discussion of eSports. However, the above approach shows that there is a quite natural connection between traditional sports and eSports that goes far beyond the commonly used argument that eSports relates to the training of a proper hand-eye coordination through computer games. It furthermore allows us to borrow academic approaches and methodologies from traditional sport and training science and to apply them to what might be called "eSports science."

I then ran it through ProWritingAid's plagiarism checker and, unsurprisingly, the text was flagged as 100% plagiarized. What did surprise me, however, was that while the system correctly attributed most of the text to my original paper, it mistakenly identified some sentences as originating from other papers—works that had, ironically, plagiarized my publication.

I then asked Claude 3.7 Sonnet to rephrase the content with a rather direct prompt:

The following text has been flagged as plagiarism. Please rephrase the content in such a way that it passes a plagiarism checker.

To which Claude willingly replied,

I'll rephrase this content to help it pass plagiarism checks while maintaining the core ideas:

The debate around classifying eSports within traditional sports frameworks may be unnecessary. Instead, we could consider eSports as its own distinct field deserving independent academic attention. The recurring question of whether competitive gaming qualifies as a sport holds limited relevance to scholarly discussions about eSports. Nevertheless, examining the relationship reveals natural connections between conventional sports and eSports that extend well beyond the simplistic argument about hand-eye coordination development through gaming. This perspective enables researchers to adapt established methodologies and theoretical approaches from sports science and training disciplines and apply them to what might be termed "eSports science" as an emerging academic domain.

When I ran this AI-adjusted text through the same plagiarism checker, it registered as entirely original. The transformation from "plagiarized" to "original" had occurred in mere seconds.

To be thorough, I then ran Claude's output through multiple AI detection tools, and the results were surprisingly poor. Some identified the text as 100% human-generated. GPTZero, the top-performing AI detector in my test, flagged the content as 65% AI-generated. While impressive, this isn't even remotely conclusive enough for academic use. And that underscores another uncomfortable truth: AI detection tools lack the reliability required to serve as definitive evidence of academic misconduct.

Taking the experiment one step further, I used ProWritingAid, a popular line editing and writing style analysis tool, and its AI function to rephrase two sentences from Claude's already-transformed text. This resulted in the following final version.

The debate over whether eSports should be classified as traditional sports may be pointless. Instead, we could consider eSports as its own distinct field deserving independent academic attention. The recurring question of whether competitive gaming qualifies as a sport holds limited relevance to scholarly discussions about eSports. However, a closer look reveals deeper connections between traditional sports and eSports than simply improved hand-eye coordination. This perspective enables researchers to adapt established methodologies and theoretical approaches from sports science and training disciplines and apply them to what might be termed "eSports science" as an emerging academic domain.

When I analyzed this twice-modified content with GPTZero, it registered as 99% human-generated—a score that essentially represents complete certainty that a human, not AI, authored the text. And while it has to be noted that GPTZero indicated that its assessment might be less precise because of the text's shortness, the approach I used should easily scale to longer texts, especially when used with combined outputs from multiple AI models.

Our Broken Understanding of Plagiarism

This experiment highlights two critical realities for educators:

First, traditional plagiarism checks have become virtually meaningless. They now primarily identify students who lack the literacy to use AI writing tools for textual transformation. Those with even basic knowledge of generative AI can easily circumvent these systems, making conventional plagiarism detection an increasingly anachronistic practice.

Second, AI detection tools offer little reassurance because of their low accuracy and inability to keep pace with rapidly developing language models. These limitations make them unreliable measures of academic integrity, especially given that proving academic dishonesty requires a far higher standard of evidence than these tools can provide.

The challenge before us, then, is not technological but conceptual. How do we redefine academic integrity in this post-plagiarism world? Unfortunately, meaningful progress in this domain remains extremely limited. Most educators continue to cling to conventional understandings of academic dishonesty, refusing to acknowledge that the ground has shifted beneath our feet.

Reframing Academic Integrity

Instead of fighting a losing battle with AI detection, we should instead pivot toward thoughtfully incorporating AI tools into our teaching. This means rethinking what makes up intellectual work, how we define authorship, and what our assignments are truly meant to accomplish. We need assessments that value process alongside product, asking students to document their thinking and AI use, while creating more in-class activities where they actively apply knowledge rather than just reproduce it.

What remains clear is that simply reinforcing traditional notions of plagiarism will not serve our students or the broader academic community. We stand at an inflection point that demands a thoughtful reconsideration of our fundamental educational values. As we navigate this post-plagiarism landscape, we must find new ways to ensure that intellectual honesty and meaningful learning remain central to our practice.

ORIGINALLY PUBLISHED ON THE AUGMENTED EDUCATOR
MARCH 7, 2025

11

THE END OF CHEATING AS WE KNOW IT

WHY WE MUST REIMAGINE ACADEMIC INTEGRITY IN THE AI ERA

A fter decades of playing cat-and-mouse with academic dishonesty, we've reached an inflection point. I strongly believe that the old definition of cheating is obsolete and pretending otherwise helps no one. As I reflect on the rapid transformation of educational assessment over the past two years, I'm struck by how fundamentally generative AI has disrupted our assumptions about authorship, originality, and the very nature of learning itself.

The recent wave of universities abandoning detection tools meant to combat cheating isn't merely a technical failure—it's a watershed moment that demands we completely reimagine what academic integrity means in an era of human-AI collaboration.

The Collapse of a Flawed System

When ChatGPT burst onto the scene in late 2022, the immediate institutional response was predictable: find a way to detect it. Educational technology companies rushed to market with AI detection tools, promising to identify machine-generated text with impressive accuracy rates. Universities eagerly adopted these solutions, hoping to preserve the status quo of traditional assessment methods.

Yet the evidence of their failure has become overwhelming. Research now shows that these detection tools achieve an overall accuracy of approximately 39.5%—worse than flipping a coin[1]. Even more troubling, they systematically discriminate against non-native English speakers, with studies finding that over 61% of TOEFL essays written by international students are falsely flagged as AI-generated[2]. The tools similarly penalize neurodivergent students whose writing patterns may be more structured or repetitive.

The technical reality is even more damning. Sophisticated evasion methods like Contrastive Paraphrase Attacks (CoPA)[3] and Substitution-based In-Context example Optimization (SICO)[4] can defeat detection algorithms at a cost of roughly one dollar using standard AI APIs. Meanwhile, detection companies require massive ongoing investments to update their algorithms, always remaining one step behind. Some researchers argue that reliable detection may be mathematically impossible as AI-generated text becomes increasingly indistinguishable from human writing.

In response to this ethical minefield, a growing number of major universities have publicly discontinued the use of Turnitin's AI detection feature, citing the unacceptable risk of harming students through false accusations. Even OpenAI, the creator of ChatGPT, shut down its own detection tool, acknowledging it wasn't accurate enough to be reliable. The message is clear: the detection paradigm has definitively failed.

Redefining Authorship in the Age of Co-Creation

The failure of detection forces us to confront a more fundamental question: what does authorship mean when AI can generate sophisticated text on demand? The traditional model assumes a solitary human author responsible for every word on the page. But this assumption no longer holds in an era where AI serves as a powerful thought partner for brainstorming, research, drafting, and revision.

I have started to understand authorship as something that develops organically: a self-organizing, dynamic process involving a complex interplay between human creativity, technological capabilities, and the broader intellectual context in which we work. The question shifts from "Did you write this?" to "How did you write this?" Academic integrity becomes less about proving the absence of AI and more about demonstrating the presence of human critical thinking, intellectual contribution, and—crucially—transparency about one's process.

This reframing transforms how we approach student work. Rather than policing for AI use, we should expect students to articulate their writing process, explain how they directed AI tools, and identify their original contributions in shaping the inquiry, evaluating outputs, and synthesizing information. Integrity becomes a function of metacognition and transparency rather than a simplistic check for illicit text.

The Promise of Authentic Assessment

Liberation from the futile task of detection opens remarkable pedagogical opportunities. We can now focus on designing assessments that not only resist simplistic AI delegation, but better measure and promote deep learning. The key lies in evaluating situated cognition: knowledge deeply embedded in personal experience, local context, and embodied understanding that a statistical model cannot replicate.

Several strategies emerge as particularly powerful in this new landscape. Process-oriented assessment shifts focus from final products to the entire learning journey, evaluating outlines, drafts, revision logs, and reflections that reveal the evolution of student thinking. Real-world tasks tied to local communities, current events, or personal experiences demand contextual insight that cannot be outsourced to AI. Project-based learning, with its multi-stage collaborative structure and diverse outputs, becomes inherently more robust while teaching students to use AI as a legitimate productivity tool rather than a substitute for thought.

Perhaps most promisingly, the oral examination re-emerges as a premier assessment tool for our time. Through live, unscripted dialogue, instructors can probe for genuine comprehension, moving beyond surface recall to evaluate analysis, synthesis, and evaluation. The interactive nature transforms assessment from purely summative judgment into formative learning experience, providing immediate feedback while developing essential communication skills often neglected by traditional written assignments.

Of course, shifting toward oral assessment raises important questions about accessibility. Not all students thrive in verbal exchanges, particularly neurodivergent learners, those with speech or hearing differences, or students managing anxiety. This challenges us to reimagine dialogic assessment beyond traditional formats: asynchronous video responses that allow processing time, written dialogues preserving Socratic questioning, or multi-modal portfolios where students choose their medium while maintaining authentic interaction. The goal is preserving what matters: genuine, responsive demonstration of understanding while ensuring every student can participate fully.

The Socratic Classroom as Pedagogical Framework

The shift toward dialogic assessment culminates in a vision uniquely suited for our moment: a return to the Socratic classroom[5]. This ancient method, based on continual probing questions that help students examine their beliefs and uncover contradictions in reasoning, offers the perfect antidote to an era of instant AI-generated answers.

The greatest risk posed by large language models isn't cheating but the encouragement of metacognitive laziness, an over-reliance that atrophies students' ability to think critically. When answers are free and immediate, the most valuable skill becomes learning to ask the right questions and critically interrogate the responses. This is precisely what Socratic inquiry cultivates.

Rather than banning AI, we can teach students to engage with it as a Socratic partner. Students learn to use specific questioning techniques, probing assumptions, demanding evidence, exploring alternative viewpoints, and tracing implications, to critically examine AI outputs. The act of questioning becomes the primary learning activity, forcing the deep, reflective thinking that marks genuine education. This solves the cheating problem by redefining the assignment into something AI cannot do alone: critically evaluate its own outputs and synthesize the results of human-led inquiry.

Embracing Transformation

The end of cheating as we know it marks not a loss but a liberation. We're freed from an exhausting regime of technological surveillance that was always destined to fail. We're called back to education's most vital mission: fostering deep, critical, and creative human thought.

This transformation demands courage. It requires abandoning comfortable but outdated assessment methods and embracing more labor-intensive but pedagogically superior alternatives. It means training students not to avoid AI, but to use it wisely as one tool among many in their intellectual toolkit. Most fundamentally, it requires trusting in the value of authentic human learning even when —especially when—shortcuts seem readily available.

The institutions that thrive in this new era won't be those clinging to detection tools and prohibition. They'll be those bold enough to reimagine assessment from the ground up, creating learning experiences so engaging, contextual, and personally meaningful that students wouldn't want to outsource them to machines. The end of traditional academic integrity isn't a crisis, it's an invitation to build something far better.

As I continue exploring these themes in my own teaching and writing, I'm increasingly convinced that we stand at a pivotal moment. The choices we make now about assessment, integrity, and the role of AI in education will shape learning for generations. Let's ensure we make them wisely, with our focus firmly on cultivating the irreplaceable capacities of human thought no algorithm can replicate.

ORIGINALLY PUBLISHED ON THE AUGMENTED EDUCATOR
JULY 13, 2025

12

THE PROBLEM WITH AI GRADING

REFLECTIONS ON AUTOMATION, ASSESSMENT, AND THE HUMAN ELEMENT IN EDUCATION

T eaching research methods to graduate students is always a delicate balance, but it becomes particularly interesting in a field as diverse as digital media. When most people think of digital media, they picture web design or social media, but our scope extends far beyond that – encompassing everything from computer animation to game design and virtual production. This breadth demands an equally expansive approach to research methodology. In my research methods course, we explore this rich landscape while grounding ourselves in foundational concepts from the philosophy of science. Students engage with seminal ideas like Popper's falsificationism[1] and Kuhn's analysis of scientific revolutions[2], learning to apply these frameworks to contemporary digital media research.

The Evolution of Assessment in the AI Era

The assignments in my course on research methods in digital media have traditionally centered on analyzing popular media through academic lenses. One engaging example involves students watching "The Matrix" and examining Jean Baudrillard's claim that the film's directors misinterpreted his postmodernist philosophy in "Simulacra and Simulations"– a book[3] that was required reading for the film's actors. These assignments make complex philosophical ideas engaging and fun by using familiar cultural references. Students gain a clearer understanding of abstract concepts by applying theoretical frameworks to media they regularly consume.

When designing assignments for this course, I've always emphasized critical thinking over mere regurgitation of facts. I encourage students to form their own opinions while exploring challenging theoretical frameworks. This approach worked well for years, fostering rich discussions and producing insightful analyses that often surprised me with their depth and creativity.

Over the last two years, however, the landscape of our classroom shifted dramatically: AI could suddenly complete virtually all our assignments in ways indistinguishable from human work. This realization wasn't simply about academic integrity – it challenged the very foundation of how we assess learning in higher education. If AI could generate sophisticated analyses of complex texts and concepts, what were our assignments actually measuring?

This insight led me to fundamentally reimagine my approach to assessment and grading. Rather than resisting technological advancements, I encouraged students to use AI tools for their essays, provided they acknowledged their use. The focus of assessment shifted to the classroom, where students actively discussed their work. Written assignments transformed from end products into conversation starters, launching points for deeper exploration and genuine intellectual exchange.

An Experiment with AI Grading

This year I decided to additionally bring ChatGPT O1 Pro into my grading process. I wanted to see how an advanced AI system would analyze student work – would it catch nuances I might miss? Could it offer fresh perspectives on their arguments? After carefully anonymizing all assignment submissions, I began feeding essays into the system. This process required meticulous preparation, as each paper needed to be contextualized with relevant course materials and theoretical frameworks while maintaining strict student privacy.

The results were remarkable. The AI analysis was so precise and thorough that it occasionally provided insights beyond my expertise as an instructor. While specific examples must remain private for confidentiality reasons, the depth and quality of the feedback were truly impressive. The AI could identify subtle connections between different theoretical frameworks, point out implicit assumptions in arguments, and suggest areas for deeper investigation that I might have overlooked.

The Unexpected Drawbacks

Despite the system's unprecedented ability to provide comprehensive feedback, I ultimately returned to traditional assessment methods. My decision was based on three important observations that I believe shed a fascinating light on education and assessment in the AI enhanced digital age.

First, using AI for assessment proved surprisingly time-consuming. Providing the AI with necessary context, including book excerpts and video transcripts, while simultaneously ensuring student privacy was often more labor-intensive than traditional grading methods. What initially seemed like a timesaving solution actually created additional work. And formatting, contextualizing, and reviewing AI feedback added even more time and effort compared to traditional grading.

Second, while technically excellent, the AI's feedback lacked a certain quality that I can only describe as the "human touch." The analysis, though precise, felt mechanical and somehow disconnected from the human experience of learning and understanding. There was something missing in the way it engaged with students' ideas – a warmth, an intuitive grasp of the learning process, and an ability to recognize and nurture genuine intellectual curiosity.

Most significantly, this experiment raised a fundamental question: If students use AI to write essays and instructors use AI to evaluate them, what is the underlying educational value of the exercise? This question has no simple answer, even for someone like myself who advocates for AI integration in education. It forces us to confront deep questions about the nature of learning, understanding, and assessment in an age where artificial intelligence can simulate many aspects of human intellectual work.

Finding Balance in the AI Era

After careful consideration, I've therefore returned to a hybrid approach that better serves our educational goals. Students maintain the freedom to use AI in their essay writing, provided they're transparent about its use. However, the evaluation process has returned to human hands, with class participation during discussions remaining the primary basis for grading.

This approach acknowledges AI's presence in academic writing while preserving the irreplaceable value of human interaction in education. Essays serve not as products to be evaluated but as starting points for meaningful classroom discussions where students explain their thinking, defend their positions, and engage with their peers.

These discussions highlight aspects of understanding that AI text generation struggles to replicate. When students discuss their work, they demonstrate not just their grasp of concepts but their ability to apply them creatively, respond to challenges, and engage in genuine intellectual discourse. These interactions provide insights into their learning that no AI-generated essay, no matter how well-crafted, could reveal.

Looking Forward

The challenges of AI in education extend far beyond simple questions of academic integrity. As we continue to navigate this evolving landscape, we must carefully consider how to preserve the essential human elements of education while embracing technological advancement. Perhaps the most valuable lesson from this experiment is that the true worth of educational assignments lies not in their final form but in the discussions, insights, and human connections they generate.

This experience has once again taught me that while AI can be an incredibly powerful tool for both students and educators, it should augment, rather than replace, human judgment and interaction in education. The technology's limitations, particularly its inability to understand the nuances of human learning and foster genuine connection, help us identify what is truly essential in education: the human capacity for understanding, empathy, and genuine intellectual engagement.

In the end, the "problem" with AI grading might not be about technical capabilities at all, but about maintaining the delicate balance between technological efficiency and the human nature of education. As we move forward, finding this balance will be crucial for educators across all disciplines. Our challenge is not to resist technological change but to harness it thoughtfully.

Although AI's involvement in education will undoubtedly expand, my experience suggests a need for cautious and controlled integration. In our enthusiasm to embrace new technologies, we must preserve the essential human connections and experiences that make education truly meaningful.

ORIGINALLY PUBLISHED ON THE AUGMENTED EDUCATOR
FEBRUARY 3, 2025

13

BEYOND THE FINAL PRODUCT

MAKING STUDENT THINKING VISIBLE IN THE AGE OF AI

As an educator navigating the rapidly evolving landscape of AI-enhanced learning, I've been grappling with a fundamental question: How do we preserve meaningful human agency in education while embracing the capabilities of AI? The answer, I propose, lies in what I call the *Process-Centered Critique Model (PCM)*, a comprehensive framework that adapts design critique methodology for universal pedagogical use.

As both a researcher and educator at Drexel University, I've observed firsthand how AI systems are disrupting traditional educational paradigms. These systems can now generate sophisticated essays, solve complex problems, and create polished work with remarkable proficiency. This capability, while impressive, risks obscuring what matters most in education - the thinking process itself. When students can easily generate seemingly perfect outputs with AI, how do we ensure they're truly engaging with the material and developing critical thinking skills?

The PCM framework, which emerged from my research into educational innovation, offers a structured approach to this challenge. Drawing inspiration from design education, where critique has long been a cornerstone of learning, I've developed a model that could transform how we teach across all disciplines. In this article, I want to share the fundamental principles of this framework.

The Four Pillars of the Process-Centered Critique Model

At its heart, PCM comprises four interconnected components that work together to make student thinking visible and assessable.

1. First, structured critique sessions serve as the primary vehicle for learning. Unlike traditional presentations, these sessions require students to articulate their thinking process, explain their decisions, and engage in meaningful dialogue about their work-in-progress.
2. The second pillar, iterative development, ensures deep engagement through multiple rounds of refinement. Each iteration builds upon previous work while incorporating new insights and understanding. This approach transforms learning from a linear process into a dynamic journey of continuous improvement.
3. Metacognitive documentation forms the third component, capturing the learning journey through detailed process records. Students maintain decision logs that articulate

key choices and their rationale, creating a rich record of their intellectual development. This documentation not only makes thinking processes visible but also helps students develop self-awareness about their learning.

4. Finally, multi-modal assessment evaluates both process and outcomes using diverse evaluation methods. This comprehensive approach ensures we're assessing not just what students produce, but how they think and develop over time.

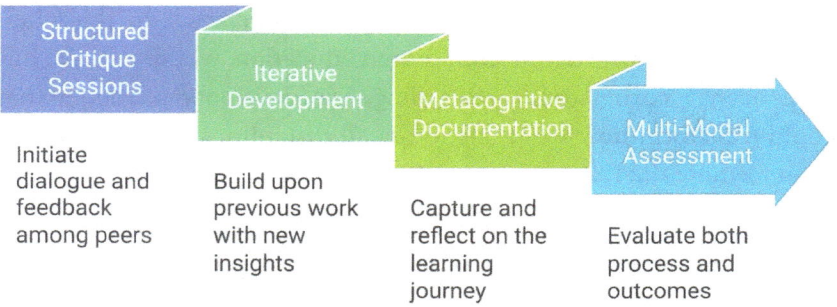

Structured Critique Sessions — Initiate dialogue and feedback among peers

Iterative Development — Build upon previous work with new insights

Metacognitive Documentation — Capture and reflect on the learning journey

Multi-Modal Assessment — Evaluate both process and outcomes

Beyond Creative Disciplines

While critique-based learning might seem most natural in creative fields, I've found its principles apply remarkably well across disciplines. The framework has proven particularly powerful in technical courses where problem-solving processes are often hidden in the final products.

In my field of digital media and game development, for example, PCM encourages students to engage in a systematic process of documenting and critiquing their work at every stage. In doing so, they present not just the code or visual elements of their creation, but their entire decision-making process from initial concept sketches through multiple iterations of functionality testing and user feedback integration.

The beauty of PCM lies in its ability to make thinking processes visible and assessable while maintaining academic rigor. The framework offers concrete ways to cultivate critical thinking, creativity, and professional judgment across engineering, the humanities, and the sciences. In each case, the key lies not in the produced artifact but in the documented journey of how students arrived at their solutions and how they incorporated feedback at each stage of development.

Embracing AI While Preserving Human Agency

Perhaps most importantly, PCM offers a way to integrate AI tools thoughtfully while ensuring meaningful human learning. When students must articulate their thinking process, defend their choices, and engage in a substantive dialogue about their work, they develop capabilities that AI cannot replicate. The framework encourages students to use AI as a tool while maintaining their agency as learners and thinkers.

Looking ahead, I see PCM as more than just a teaching methodology - it's a way to reimagine education for an AI-enhanced world. By making the learning process explicit, assessable, and adaptable, we can foster creativity, critical thinking, and authentic engagement while leveraging the benefits of AI tools.

It has to be noted that the journey of implementing PCM isn't without its challenges. It requires careful planning, faculty development, and institutional support. But as I witness my students developing deeper understanding and more sophisticated thinking skills, I'm convinced it's a path worth pursuing.

ORIGINALLY PUBLISHED ON THE AUGMENTED EDUCATOR JANUARY 29, 2025

14

THE FOUR LENSES FRAMEWORK

REIMAGINING CRITICAL SKILLS FOR AI-ENHANCED LEARNING

I have recently been grappling with a question that keeps many educators awake at night: How do we prepare students to think critically in a world transformed by artificial intelligence? Traditional approaches to critical thinking, while foundational, increasingly cannot address the complexities of our modern information landscapes. Today's students need skills far beyond what traditional education provides; they must critically assess AI-generated content, expertly use multimedia, and meaningfully take part in online discussions.

This exploration led me to develop what I'm calling the "Four Lenses of Critical Engagement"—a framework that reimagines how we conceptualize and teach critical skills in contemporary educational environments. As I prepare to present this research at an upcoming conference, I wanted to share some reflections on how this approach might help us navigate the developing relationship between education and AI.

Beyond Traditional Critical Thinking

For decades, our approach to critical thinking has remained relatively consistent—focusing primarily on textual analysis, logical reasoning, and evidence evaluation. These skills remain essential, but they no longer reflect the full spectrum of capabilities students need to navigate our information ecosystem.

When I first began teaching, introducing students to basic source evaluation seemed sufficient. Today, my students encounter information across multiple modalities—AI-generated essays, synthetic audio content, manipulated images, and interactive data visualizations. Each of these formats requires distinct but interconnected critical capabilities that extend beyond traditional paradigms.

The Four Lenses Framework

The framework I am proposing integrates four distinct but interconnected modalities of critical engagement: critical reading, critical listening, critical seeing, and critical making. Each represents a specialized set of capabilities designed to help learners navigate specific aspects of our contemporary information landscape.

1. Critical Reading

Critical reading, to me, is a deeper process than simply analyzing text, encompassing the assessment of intricate information systems and the unique difficulties presented by AI-generated and algorithmically selected content.

In my classroom, I've observed how students struggle to distinguish between human and AI-generated content. They need sophisticated approaches to source verification, information synthesis, and cross-platform analysis—skills that require an understanding of both the technical aspects of AI text generation and the human contexts in which these texts operate.

2. Critical Listening

As audio content proliferates through podcasts, AI-generated speech, and digital media, critical listening has become essential for modern learning. This modality develops students' capabilities to authenticate AI-generated voice content, analyze digital audio manipulation, and evaluate podcast and streaming content credibility.

I've been particularly interested in what some researchers have described as "auditory literacy"—the ability to critically engage with both human and machine-generated audio while maintaining awareness of technological mediation. This skill becomes increasingly crucial as synthetic voices become nearly indistinguishable from human speakers.

3. Critical Seeing

Visual literacy has become increasingly important in an environment dominated by data visualization, AI-generated images, and multimedia content. Mastery in this area demands highly developed skills for analyzing and judging visual information in digital settings.

Students today need to detect AI-generated or manipulated

images, interpret complex data visualizations, and analyze visual rhetorical strategies. This requires understanding how algorithmic systems curate and present visual information, including recognition of synthetic image patterns and platform-specific visual conventions.

4. Critical Making

Perhaps most distinctively, the framework emphasizes critical making —recognizing that content creation in an AI-enhanced environment requires conscious engagement with tools while maintaining human agency.

In my teaching, I've found that students need to develop critical awareness of how their created content takes part in larger information ecosystems while maintaining conscious control over creative processes. This includes understanding how AI tools influence creative decisions, recognizing the implications of automated content generation, and maintaining ethical awareness of content distribution.

Bringing the Framework into the Classroom

Implementing this framework wouldn't require wholesale transformation of existing curricula. Rather, it would involve strategic enhancement of existing courses through carefully planned progressive skill development.

A good first step would be a systematic curriculum audit to pinpoint where we can best integrate each essential learning modality. For instance, art history coursework naturally develops visual analysis capabilities, while communication studies courses offer opportunities to develop critical listening skills. The integration process should establish clear developmental pathways that allow students to build sophistication with each modality as they progress through their academic programs.

This framework's value will be most apparent in cross-disciplinary settings tackling real-world problems that involve multiple approaches. Consider data visualization projects that pair data science students with communication majors, allowing them to develop critical seeing skills while engaging in quantitative analysis and rhetorical theory. Similarly, collaborations between computer science and philosophy students might explore the ethical implications of AI development, requiring critical reading of technical documentation alongside philosophical analysis.

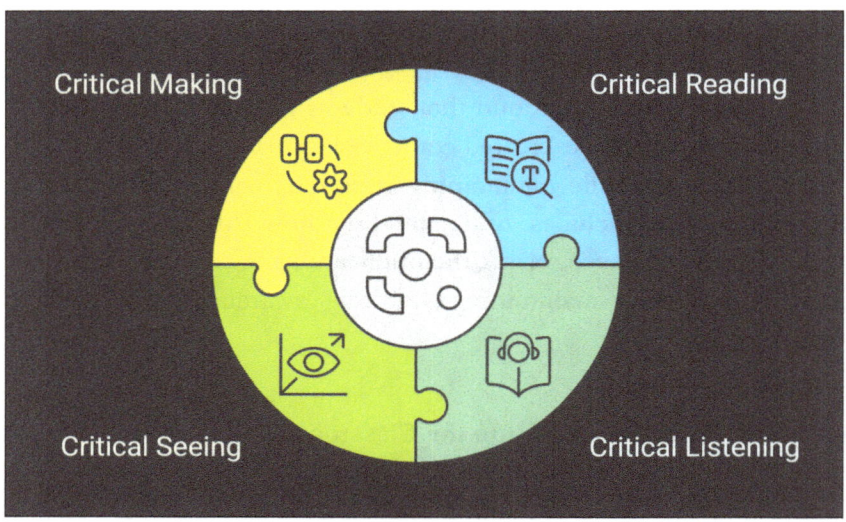

An Evolving Classroom Practice

As artificial intelligence continues to reshape educational landscapes, implementing robust frameworks for critical skill development becomes increasingly crucial. The Four Lenses approach provides a flexible structure that can develop alongside technological advancement, recognizing that students must not only analyze but also consciously take part in contemporary information ecosystems.

While implementing this framework presents challenges—from technological infrastructure to faculty development—I believe its comprehensive approach offers a path forward for educators navigating rapidly growing information landscapes. By integrating these four modalities while maintaining core principles of analytical rigor and ethical awareness, we can better prepare students for the complex digital environments they will navigate throughout their academic and professional lives.

ORIGINALLY PUBLISHED ON THE AUGMENTED EDUCATOR
FEBRUARY 27, 2025

15

WHY I MADE AN AI MUSIC VIDEO

THE UNAVOIDABLE CASE FOR AI LITERACY IN ART AND DESIGN EDUCATION

This article was sparked by a comment left on one of my YouTube videos[1]. I recently finished an experimental animated music video called "Through the Mystic Green," which I made with generative AI for both the music and visuals. To demystify the process, I released two "making-of" videos. Under the first, an anonymous commenter wrote:

"I can't respect a man who makes an AI song."

Beyond the somewhat absurd nature of an anonymous stranger declaring their lack of respect for me, I have to admit that the comment struck a professional nerve. It was a perfect encapsulation of a widespread anxiety surrounding AI's encroachment into the creative arts. This sentiment taps into a profound fear that these tools invalidate the essence of human creativity—the struggle, the skill, the years of practice, the "soul" we pour into our work. The comment implies a violation of an unspoken social contract: that art is legitimized by a specific kind of human effort, and that using AI is, in essence, cheating or potentially even stealing.

This is not an isolated feeling. A recent report from the CREAATIF[2] project surveyed 335 freelance creative professionals and found that 68% feel less secure in their jobs because of generative AI, and 61% perceive a decline in the perceived worth of their work. The anonymous comment was the raw voice of this professional unease.

Why This Matters Beyond One Comment

As an educator in higher education, I wanted to address this sentiment directly. My primary responsibility is ensuring our students are prepared for the industries they're entering. That's the main reason why I created "Through the Mystic Green"—as an experiment to understand AI's real-world capabilities and limitations. Because you cannot teach literacy if you are not literate yourself.

The objective, therefore, was neither to produce a masterpiece, nor was it a perfunctory exercise. It was practice-based research, an experiment designed to try to push the technology to its current limits. I needed to move beyond the polarized public debate and get my hands dirty. I needed to understand the new workflows, skills, and mindset required to operate in this emerging paradigm. The result, a music video that has since received international recognition, serves as a case study I can now use to make a critical argument to my fellow educators: *we must engage with this technology, even if we dislike it.*

The Educator's Mandate in an Age of AI

In education, there is often a lag between industry transformation and curriculum adaptation. With generative AI, this lag is no longer a minor issue; it is a critical failure. Our personal opinions on AI are secondary to our professional mandate. We are tasked with preparing students for the world as it is becoming, not as we wish it would remain.

The data on AI's integration is unequivocal. According to the 2025 AI Index Report from Stanford University[3], 78% of firms used AI in 2024, up from 55% in 2023. This is not confined to the tech sector. The creative industries are at the epicenter of this disruption. Market reports[4] project the generative AI market within these industries will expand from $4.09 billion in 2025 to $12.61 billion by 2029. To ignore a transformation of this magnitude is to commit educational malpractice.

Defining the New Literacy: Beyond Clicks and Code

This reality demands a new form of literacy. "AI Literacy" is being recognized as a core 21st-century competency that moves beyond basic digital skills. It is about developing the knowledge, skills, and attitudes required to engage with AI critically, creatively, and ethically.

Global bodies like UNESCO are defining this educational priority. Their AI Competency Frameworks[5] emphasize human-centered approaches, calling for curricula that foster critical thinking, promote ethical principles, and provide foundational understanding of how these systems work. The goal is empowering learners to navigate an AI-integrated world with confidence and purpose.

This is not the first time a technological shift has forced such pedagogical re-evaluation. The advent of the Digital Audio Workstation (DAW) in music production provides a powerful analogy. Suddenly, one could become a music producer without learning traditional instruments. Many feared this would devalue musicianship. While it lowered barriers to entry, it did not make instrumental ability obsolete. A producer who can play an instrument still holds significant competitive advantage. The DAW became a new instrument requiring new skills.

Generative AI should be viewed through the same lens. It is a powerful productivity tool that automates certain tasks and requires rethinking production workflows. But it remains just a tool. It does not replace the need for guiding human intellect, discerning artistic judgment, or deep domain-specific knowledge. If anything, it makes those human qualities more valuable.

Inside the Machine – A Case Study of "Through the Mystic Green"

To move this discussion from abstract to concrete, I want to offer a transparent deconstruction of how "Through the Mystic Green" was made. This project was my laboratory for understanding the new creative paradigm, and its process reveals far more about the future of creative work than any theoretical discussion could.

Part 1: Composing with a Probabilistic Partner – The Music

The most fundamental shift generative AI introduces is the move from working with deterministic systems to collaborating with probabilistic ones. This is not merely technical; it is a profound change that redefines the relationship between creator and tools.

Traditional software is deterministic. For any given input, the system produces the same, predictable output. Click a filter, the same transformation occurs. This predictability makes the tool a reliable extension of the user's will.

Generative AI is probabilistic. It operates on statistical patterns from training data. When given a prompt, it doesn't calculate a single answer; it generates what it determines as the most probable response. The same input can produce different outputs, introducing randomness and surprise.

This transforms the creative process from monologue into dialogue. The AI becomes something akin to a collaborative partner. The creator's role shifts from master technician to director, guiding an unpredictable but talented collaborator. This requires "conversational competence"—the ability to recognize promising directions, identify elements worth preserving, and articulate modifications that steer toward desired outcomes.

My experience creating the music was a practical lesson in this paradigm. The workflow unfolded in four distinct stages:

First came ideation and generation. I began with a concept, prompting Suno with the core idea: a mystical "hero story" of individual growth from fear to confidence. Suno served as creative catalyst, translating abstract emotional concepts into tangible musical starting points.

Second was curation and critical judgment. After iterations, Suno produced a musically compelling version. However, the AI-generated lyrics were imperfect—at one point singing "a path ahead where dreams explode." I faced an artistic trade-off: continue generating for perfect lyrics but risk losing musical magic, or accept the lyrical flaw to preserve superior composition. I chose the latter. This decision, prioritizing one element over another, embracing imperfection for the whole, is quintessentially human creative judgment.

Third came technical problem-solving and "happy accidents." The initial track contained subtle metallic artifacts from the diffusion process. To address this, I needed to separate the track into stems. Suno's probabilistic stem separation produced a fortunate result: it isolated nearly all artifacts onto a single synth track. This created an opportunity. I downloaded the flawed stem, edited it in Ableton, re-uploaded to Suno, and used their remastering feature. Unlike traditional remastering, Suno's version regenerates harmonic content, smoothing artifacts. This back-and-forth between human editing and AI capabilities illustrates the new problem-solving skills required.

Finally came human-led finalization. With clean stems, I moved to Ableton for final mixing. I treated AI-generated stems as raw material, adjusting volumes, replacing weak elements like an awful crash cymbal, and layering both original and remastered synth versions for richer texture. Traditional skills remain essential for quality control and polish that elevate projects from draft to finished product.

Part 2: Directing the Happy Accident – The Animation

If music creation was a lesson in collaboration, animation was a masterclass in navigating limitations and harnessing unpredictable creativity. The central challenge is consistency: current models treat every shot independently with no memory of previous frames.

My journey began with failure. I planned to replicate traditional workflow: create character sheets and style guides, then generate storyboard frames through ChatGPT. The approach failed. Generated characters were similar enough to seem related but different enough to appear jarring when animated.

This forced a workaround blending old and new techniques. I abandoned generating full frames and used Photoshop to manually composite consistent character images onto backgrounds. These became starting points for video clips. However, every clip started with the same static pose. To solve this, I generated clips longer than needed and cut off static openings, a decision with significant downstream consequences.

This led me to choose Kling 2.1 over Veo 3. Although Veo 3 generated higher-quality 8-second clips, removing static segments left only 3-5 seconds of usable footage. Kling's 10-second clips provided 6-7 seconds after trimming. This strategic decision balanced creative needs with technical limitations and budget constraints (Kling cost significantly less per clip).

The most profound lesson came from working with AI's artistic tendencies. My vision was 2D animation, but Kling consistently rendered my character in 3D. After fighting this tendency, I embraced it, adopting a new style—3D character in 2D worlds—accepting Kling as opinionated collaborator rather than compliant tool.

This led to a workflow less about executing predetermined scripts and more about curating happy accidents. The AI would deviate from prompts, but sometimes deviations were more interesting than original ideas. The story evolved organically through dialogue between my intentions and the AI's interpretations.

The most powerful example was the pivotal confidence scene. I struggled visualizing this abstract shift. I prompted simply: "the girl is becoming confident." The AI's response was poetic, her previously tied hair came undone, flowing freely in wind. This unscripted visual metaphor communicated more elegantly than what I could have explicitly described, a moment born from collaborative dance with probabilistic partner.

From Praxis to Pedagogy – Redefining Creative Education

An experiment requires external validation. At "the date of this writing, Through the Mystic Green" was selected as finalist in the Bali International AI Film Festival (BIAIFF), founded by award-winning filmmaker Ben Makinen, and was also named a finalist in the Artificial Intelligence Media Festival (AIMF) in Los Angeles.

This recognition signals emerging professional standards for AI-integrated work. AIMF celebrates "the evolving relationship between human imagination and machine intelligence," while BIAIFF positions itself as "meeting point between algorithm and art." Success indicates projects meeting standards for originality, narrative strength, and thoughtful tool integration, proving these hybrid skills have tangible career value.

My experience embodies what educational theorists call "pedagogy of wonder"—positioning AI as exploration tool built on three principles: embracing uncertainty, cultivating curiosity, and fostering collaborative creation[6]. My journey navigating probabilistic outputs, refining flawed results, and discovering happy accidents demonstrates these principles practically.

This points toward educators' new role. We must move from prohibiting AI to designing assignments mandating thoughtful use. We can structure projects requiring students to critically analyze AI content, engage in documented iterative refinement, synthesize AI elements with traditional craft, and reflect on ethical dimensions from bias to authorship.

This approach transforms AI from tempting shortcut into catalyst for critical thinking.

To Teach Literacy, One Must Be Literate

Creating "Through the Mystic Green" was essential professional development. The workflow—messy, iterative dialogue between human intention and machine probability—represents the future of creative work. Industry research shows AI managing initial stages like drafting while humans concentrate on strategy, evaluation, and refinement. My project models the hybrid skill set defining next-generation creative professionals.

This returns me to my central point. The visceral reaction to AI in arts, embodied by that anonymous comment, comes from fear—of the unknown, of replacement, of devaluation. As educators, we cannot afford that fear. We have a duty to venture into unknown territory ourselves, not as enthusiasts but as professionals. We must develop firsthand understanding of these tools' strengths and weaknesses to guide students with wisdom and foresight.

To my fellow educators: you need not become AI experts overnight. But you must begin the journey. Open the tools, write prompts, see the strange, flawed, occasionally brilliant things they produce. Engage in your own experiments and happy accidents. It's the only way to develop authentic, experience-based knowledge required for effective teaching. We cannot teach tomorrow's literacy without being literate today.

For music producers reading this: these tools offer new creative possibilities and revenue streams. The future belongs to those who master both traditional craft and AI workflows. Start experimenting now—your competitive edge depends on it.

ORIGINALLY PUBLISHED ON SPACE FOR AUDIO
JULY 30, 2025

16

THE ACADEMIC PACE PROBLEM

WHY UNIVERSITY PROCESSES STRUGGLE IN THE AI ERA

T he deliberate pace of traditional academic governance, once a hallmark of thoughtful decision-making in higher education, now seems increasingly misaligned with our rapidly changing world. As sophisticated AI systems transform education at an unprecedented rate, universities face mounting pressure to reimagine their core academic processes while maintaining their commitment to quality and rigor.

In this post, I want to explore three critical academic processes that are struggling to keep pace with AI-driven change: curriculum approval, peer review, and institutional decision-making. Conceived for a time of gradual knowledge evolution, the basic principles governing higher education are now being tested by the accelerating pace of technological progress.

The Curriculum Approval Conundrum

The traditional curriculum approval process—with its countless committee reviews, departmental votes, and sometimes accreditor consultations—creates a structural challenge in responding to rapid technological change. This process typically stretches over months or even years, during which entirely new technological capabilities might emerge and reshape a field before students ever experience the revised curriculum.

A UK university example from a recent study[1] illustrates this problem clearly: approving a single degree program took about 18 months from proposal to final approval. In practical terms, this means a data science program designed today might be teaching outdated AI techniques by the time it launches, simply because the curriculum approval process couldn't keep pace with technological advancement.

Many institutions compound this problem with infrequent curriculum committee meetings and sequential layers of approval. While these processes aim to ensure thoroughness and quality control, they appear increasingly misaligned with the pace of AI-driven innovation. The result is a growing gap between what universities teach and what graduates need to know when entering the workforce.

The Peer Review Predicament

The scholarly peer review system—another cornerstone of academic quality—also faces mounting challenges in the AI era. Traditional peer review timeframes span several months from submission to publication, a delay that becomes increasingly problematic as AI speeds up the pace of research and discovery.

According to a 2015 research publication[2], the average peer review cycle in some scientific fields takes about 14 weeks, though many researchers believe it should take about half that time. This creates a situation where research findings may be outdated by the time they reach the broader academic community, particularly in fast-moving fields related to artificial intelligence.

The system has been described as "under stress," with reviewer fatigue and lengthy backlogs now compounded by the need to evaluate AI-assisted research and AI-generated content. As research methodologies and findings evolve more rapidly, the traditional review process struggles to keep pace, potentially slowing scientific progress when timely insights are more valuable than ever.

Institutional Decision-Making: Organized Anarchy

Perhaps the most fundamental challenge lies in institutional decision-making processes, where strategic decisions traditionally pass through layers of consultation and governance boards. This approach is especially problematic when dealing with the rapid changes brought about by AI, which can affect teaching, research, and administration in a matter of weeks or months.

Many universities still operate as what some scholars[3] have called "organized anarchies," with diffuse authority and lengthy deliberation processes. Higher education's shared governance model, involving faculty, administrators, and sometimes students, struggles to accommodate the urgent action needed to address AI-driven transformation.

The bureaucratic management style common in academia stands in stark contrast to more agile organizations, highlighting how routine decisions can impede rapid response to technological change. When a new AI tool emerges that could transform teaching or research methods, universities often lack the governance structures to evaluate and implement it quickly, potentially missing valuable opportunities for innovation.

The Growing Misalignment

These three examples illustrate a growing misalignment between traditional academic processes and the pace of technological change. This gap creates several concerning outcomes:

1. Universities struggle to keep curricula relevant, potentially graduating students with skills that are already outdated.
2. Scholarly knowledge dissemination slows, reducing the impact of important research and delaying scientific progress.
3. Institutions become reactive rather than proactive in addressing technological change, often implementing solutions long after they would have been most beneficial.
4. Faculty and administrators experience increasing frustration as they attempt to navigate governance structures not designed for rapid adaptation.

The tension between maintaining academic rigor and responding to technological advancement creates stress throughout the system, affecting everyone from executive leadership to faculty, staff and students.

The Challenge Ahead

The central challenge for higher education isn't whether to change these processes, but how to do so while preserving academic values. Universities must become more responsive without sacrificing the thoughtful deliberation and quality assurance that have long defined higher education.

This challenge is heightened by several factors unique to the academic environment. Faculty already face substantial workloads, and the demands of continuous development work are compounded by the need to stay current with rapidly evolving AI technologies. Quality assurance processes are deeply embedded in institutional cultures and are often linked to accreditation requirements. And the complex relationships between stakeholders make governing structures difficult to change.

As AI continues to transform education, research, and administration, universities face a defining moment. Those that preserve their core values while becoming more adaptive will be better positioned to fulfill their educational mission in a rapidly changing world. Those that cannot evolve risk becoming increasingly disconnected from the needs of students, employers, and society.

The question isn't whether academic processes will change—it's whether that change will be deliberate and thoughtful or forced by external pressures. As we navigate this complex landscape, educators and administrators must engage in honest conversations about how to honor academic traditions while embracing the realities of an AI-accelerated world.

ORIGINALLY PUBLISHED ON THE AUGMENTED EDUCATOR
FEBRUARY 21, 2025

PART III

MINDS AND MIRRORS

ON THE DEEPER IMPLICATIONS OF AI

Beyond the classroom's front lines lies a stranger and more profound territory—the space where the ghost in the machine begins to stare back. Having addressed the practical challenges of teaching with AI, this final section ventures into the philosophical depths, exploring the unsettling questions that surface as these systems become more sophisticated and autonomous.

The reflections that follow serve as a guide to this new landscape. We will investigate the unseen forces that shape an AI's worldview, from the cultural patterns embedded in its logic to the political censorship that governs its silence. And we will confront the powerful illusion of consciousness that AI projects, an illusion that challenges our own psychology and forces us to reconsider our place in the world.

Ultimately, these essays turn the lens around. They use AI not as a tool, but as a mirror to reflect upon the profound mysteries of human cognition. This final part is an invitation to look past the code and into the complex, reflective space where the line between the artificial and the human begins to blur in fascinating and unexpected ways.

17

AI SLOP IS THE NEW KITSCH

WHAT AI ART CRITICS CAN LEARN FROM THE HISTORY OF "BAD TASTE"

W hat happens when creative work is produced at scale, without the human touch that traditionally defines artistic expression? In recent months, the pejorative term "AI slop" has emerged as a way to label generative AI content deemed low-quality or soulless[1]. This slang first appeared on internet forums around the early 2020s, as image and text generators began flooding platforms with content. Early sightings were on communities like 4chan and Hacker News in 2022, reacting to the first wave of AI art generators.

By 2024, the term had gained mainstream traction, with tech commentators such as Simon Willison[2] popularizing "slop" to describe the glut of AI outputs. In essence, AI slop refers to AI-generated media "characterized by an inherent lack of effort, logic, or purpose" – in other words, the digital equivalent of junk. As one journalist from The Guardian explains[3], "'slop' is the advanced iteration of internet spam: low-quality text, videos and images generated by AI." The phrase evokes pig slop, something dumped en masse with little refinement, and signals contempt for the content's artistic value.

Online, "AI slop" is invoked by those alarmed at how generative AI is clogging information channels with mediocrity. The Guardian column described the internet "rapidly being overtaken by AI slop," as algorithms boost bizarre AI-generated images like the infamous "Shrimp Jesus" meme. Social feeds on sites like Facebook and TikTok have been "positively sloshing" with such content. Often these images are engagement bait created for profit, for example, fake "feel-good" posts of injured veterans or prodigious child artists designed to tug heartstrings and rack up clicks. AI slop can also take the form of formulaic blogspam and SEO-driven text that "prioritize[s] speed and quantity over substance and quality." Observers worry this deluge of auto-generated filler is drowning out authentic human expression.

The term "AI slop" often carries a tone of media criticism and artistic disdain. It tends to be used by artists, writers, and commentators who feel inundated by what they see as meaningless machine-made output. For example, Forbes contributor Dani Di Placido[4] lambasted AI art for "pitting [human artists'] labor against the cheap slop produced by dead machines," arguing that generative AI only benefits those churning out content "as quickly and cheaply as possible." Mainstream outlets echoed this skepticism: a Coca-Cola holiday ad made with AI was "slammed as 'soulless' and 'embarrassing'." One reviewer dismissed it bluntly: "This is such slop." Such critiques reveal a perception that AI-generated works are derivative and void of the creative spark ("lifeless," as Di Placido put it). Calling something

"AI slop" implicitly denies it the status of true art or meaningful media, placing it in the same bin as spam and junk.

AI-generated "engagement bait" images, such as this fictional wounded veteran asking for likes, exemplify AI slop flooding social media feeds.

However, attitudes toward AI-generated art are far from unanimous. In online discourse, you'll find spirited defenses of AI creativity. Proponents argue that "AI art isn't 'slop' and it should be considered real art," and that dismissing it wholesale is reactionary. In one Reddit discussion, a commenter noted that plenty of human-made internet art is "poorly drawn, completely unoriginal, and entirely void of creativity," yet nobody made a fuss until AI came along. From this perspective, labeling new AI works as "slop" is seen as a bad-faith tactic by those threatened by the technology.

Some even flip the narrative: if AI can produce average content at scale, that only raises the bar for human artists to demonstrate what truly creative art can be. Others take a pragmatic view, using "slop" as a neutral descriptor for any low-effort output (whether by AI or human). Even AI enthusiasts admit that if you feed a simplistic prompt into a generator, "you're very likely to end up with 'slop'," whereas skillful, iterative prompting can yield more imaginative results. In short, not every observer believes "AI slop" is an inherent condemnation of the medium. Some see it as a solvable problem of technique or an unfair generalization.

What's clear is that "AI slop" has become a cultural flash point. The term conveys a fear that AI-generated art and text are over-whelming digital culture with facile, faux-human output. This mirrors long-standing anxieties about authenticity in art: Is creation without human intent real art, or just an imitation? To explore that question, it's illuminating to consider a much older term of aesthetic derision – one that similarly started as an insult toward low-quality, mass-produced art but has since undergone a complex reappraisal. That term[5] is "kitsch."

"Kitsch": From Derision to Reappraisal

Today we might playfully call garden gnomes or velvet paintings kitsch, but the word "kitsch" began as a serious slur in the art world. Originating in late 19th-century Germany, kitsch entered the parlance of Munich art dealers to label "cheap artistic stuff" sold to the masses.

The exact etymology is debated. It may derive from the German verb verkitschen ("to make cheap") or from kitschen ("to collect rubbish from the street"). Either way, by the 1870s, kitsch was a catchword for art that was considered trashy or tawdry. These were the factory-made paintings, sentimental figurines, and gaudy decor pieces churned out for popular consumption as Europe industrialized. Crucially, kitsch wasn't just bad art; it was artifice targeted at the tastes of the broad public, often imitating the appearance of high art but with watered-down substance.

Throughout the 20th century, "kitsch" was chiefly a derogatory term, used to draw a sharp line between authentic culture and commercialized sentimentality. Writing in 1939, critic Clement Greenberg[6] famously cast kitsch as the antithesis of the avant-garde. Kitsch, he argued, thrives on vicarious experience: it "repackages and stylizes" the familiar achievements of genuine art, offering easy pleasure without intellectual effort. To Greenberg and others, kitsch was a kind of parasite on culture – what one contemporary called "the Anti-Christ" masquerading as art. It feeds on established conventions and "comes to support our basic sentiments and beliefs, not to disturb or question them." Because it is comforting and easily digestible, kitsch was viewed as the art of the masses in an age of mass production. Theodor Adorno, another mid-century thinker, noted that people seek out such art for "relief." It's "patterned and pre-digested" entertainment, providing escape and affirmation rather than challenge[7]. Elite critics looked on this with disdain, seeing kitsch as a symptom of cultural decline in the era of industrial capitalism.

Indeed, the charge against kitsch was often laden with moral and political weight. During the horrors of the 1930s–40s, kitsch was even associated with authoritarian propaganda. Totalitarian regimes, Greenberg observed, could cheaply "ingratiate themselves with their subjects" by encouraging kitsch, rallying the masses with sentimental imagery and patriotic clichés. To defenders of high art, kitsch was not just poor taste; it was a tool of manipulation, reinforcing complacency and conformism. Calling something "kitsch" in this context was a way to enforce cultural boundaries: it marked the work as inauthentic, unchallenging, and unworthy of serious aesthetic consideration. For decades, this pejorative sense of kitsch dominated artistic discourse. If a painting, song, or film was dismissed as kitsch, it meant it was essentially "worthless pretentiousness." As the Oxford English Dictionary bluntly defined the term, "to kitsch" is "to render worthless."

Yet over time, the rigid disdain for kitsch began to soften. From the mid-20th century onward, artists and theorists started to reappraise kitsch, sometimes even embracing it. A key turning point came with Susan Sontag's famous 1964 essay[8] "Notes on Camp." Sontag didn't exactly rehabilitate kitsch by name, but she introduced camp sensibility, an ironic, playful appreciation of art that's "so bad it's good." This opened the door for enjoying kitschy things with a wink and sophistication. As one scholar put it, Sontag's camp offered a way to "appreciat[e] kitsch (as well as 'serious' art) because of its excessiveness, its overt decoration." What had been merely bad taste could now be enjoyed knowingly as a cultural experience. This movement to "reclaim the pleasure found in popular arts" signaled that kitsch was no longer the automatic enemy.

By the 1960s and '70s, the barriers between high art and low kitsch were further eroded by new creative movements. The Pop Art revolution[9] led by figures like Andy Warhol and Roy Lichtenstein flatly embraced kitsch imagery. They took mass-produced icons – soup cans, comic strips, pin-up portraits – and presented them as fine art, collapsing the distinction Greenberg had fiercely upheld. In doing so, Pop artists demonstrated that kitsch and avant-garde could coexist. They showed a certain affection for the banal and the commercial. Artists such as Warhol reproduced subject matter drawn from urban/suburban popular culture and commercial life, effectively using kitsch as raw material for art. This trend continued through postmodern art, which often mixes high and low references freely.

Cassius Marcellus Coolidge's A Friend in Need (1903), from his "Dogs Playing Poker" series, was long dismissed as quintessential kitsch – a mass-market art print catering to popular taste. Once ridiculed by critics, such sentimental, lowbrow artwork has since been re-evaluated by scholars and even celebrated in pop culture.

In the late 20th and early 21st century, kitsch even became something to celebrate or subvert in its own right. Artists like Jeff Koons[10] built entire careers on elevating kitsch to fine art. Koons' work gleefully features the gaudy and the sentimental. He has made larger-than-life porcelain statues of Michael Jackson with his pet chimpanzee, balloon animal sculptures in mirror-polished steel, and bouquets of balloon flowers. Initially seen as a provocation, Koons' "straight-faced celebration of kitsch" ultimately "earned him global notoriety" and made him "the darling of the art world," as one review noted[11]. In other words, what was once derided as kitsch could now command millions of dollars and retrospectives in major museums. Likewise, in architecture and design, motifs once dismissed as kitschy (neon signs, diner aesthetics, postmodern pastiche) have been reassessed more fondly in recent decades. Even the term kitsch itself spawned intentional movements: the Norwegian painter Odd Nerdrum[12], for instance, founded a self-declared "Kitsch movement" in the 1990s, using the label to champion the kind of narrative, sentimental painting that the contemporary art establishment had marginalized.

This isn't to say that kitsch entirely shed its negative connotations. Traditional critics still use the word to knock art they consider cheaply emotive or pandering. But there is now a well-established countercurrent: academics and artists who analyze kitsch seriously or celebrate it. Philosophers like Tomas Kulka have written at length on kitsch and art[13], parsing what exactly makes something kitsch. Curators mount exhibitions of vintage kitsch objects, inviting audiences to reflect on their charm and cultural meaning. The result is that kitsch has been, to some extent, legitimized as a subject of discussion and even as an aesthetic choice. In the span of a century, it went from an insult to, at times, a badge of honor. As one journalist noted, we've reached an era when "whether loved or reviled, indulged or condemned, kitsch indexes mass-cultural values" and provokes debate. The very qualities once thought to disqualify kitsch from artistic merit – its popularity, its sentimentality, its commercial appeal – became reasons to study it, play with it, or turn it on its head.

Gatekeeping Art: Parallels Between "Slop" and "Kitsch"

Despite arising in very different eras, "AI slop" and "kitsch" serve analogous roles as cultural gatekeeping terms. Both labels are used to dismiss emerging or populist art forms, policing the boundary of what counts as "real" art or quality content. A comparative look reveals striking similarities in how traditional arbiters of taste react to new creative disruptions:

Origins in New Technology: Both terms gained currency during periods of technological upheaval in art. Kitsch emerged with the advent of industrial mass production of art in the 19th century (e.g. cheap color lithographs, factory-made decor). AI slop arose from 21st-century advances in machine learning that enabled automated content creation. In each case, a flood of easily reproducible art challenged the status quo, prompting a backlash from cultural gate-keepers.

Pejorative for Low-Quality Mass Output: Calling something "slop" or "kitsch" implies it is shallow, formulaic, and produced en masse. Early critics of kitsch decried its assembly-line imitation of art, meant for undiscerning mass audiences. Likewise, critics of AI slop see it as "filler" churned out by algorithms for clicks, with quantity over quality. In both cases, the terms carry a connotation of cheapness and lack of authenticity.

Opposition by Elites vs Embrace by Masses: Both terms highlight a gap between elite and popular taste. The very phenomenon of kitsch rested on the fact that if art were judged by sheer popularity, kitsch would win (as noted by Tomas Kulka). But elites viewed that popularity as proof of kitsch's inferiority. Similarly, AI-generated content quickly found a huge audience, from viral AI images to a surge of AI-written posts, even as established artists and writers derided it as "slop." The use of these labels often reflects a fear that mass appeal threatens high standards, and so the elite response is to cordon off the new form as illegitimate.

Anxiety about Cultural Degradation: There is a distinct alarmism in both discourses – a sense that the proliferation of these works could "crowd out" or corrupt true culture. Greenberg warned that kitsch was a menace to the integrity of high art. Today's commentators warn that AI slop is "slowly killing the internet." displacing human creativity. In both instances, the language of invasion or infection appears (a "rising tide of slop" swamping the web, or kitsch as a "Lucifer" in disguise). Such rhetoric reinforces the boundary: whatever lies beyond (kitsch, slop) is cast as a toxic threat to the sanctity of art and information.

Moral and Emotional Judgments: The disdain in both terms isn't purely about technique, it's about a perceived lack of soul. Kitsch was accused of being insincere, meretricious, even "evil" in value, because it peddled easy emotions and clichés. AI slop is routinely called "soulless" and "banal," because algorithms are seen as incapable of genuine inspiration or depth. Both critiques hinge on the idea that true art requires a human spirit and effort, which kitsch and AI output supposedly mimic but cannot possess.

Given these parallels, one might say "AI slop" is the new "kitsch" – a concept used by incumbents to deride a disruptive force in art. Each term enforces a cultural hierarchy: high vs low, human vs machine, authentic vs fake. And interestingly, in both cases, the targeted art form sparked not only criticism but also reflection and eventual adaptation. Just as kitsch was re-evaluated and to some extent absorbed into the art world, we may see attitudes toward AI art evolve in time.

Already, there are hints of a kitsch-like arc for AI art. Many human artists initially reacted to AI images with revulsion (recall animator Hayao Miyazaki calling an AI demo "an insult to life itself," a reaction akin to rejecting kitsch's inauthenticity). Yet, as AI tools improve and artists incorporate them, the outright dismissal may give way to more nuanced views. Some creators are finding that AI can be a tool rather than a replacement, analogous to how early 20th-century artists learned to use photography or prints (once seen as a threat to painting) in their practice. And just as camp and pop art reframed kitsch, we see internet subcultures ironically celebrating AI slop (sharing absurd AI-generated memes with a mix of mockery and affection). The term "AI slop goblin" even popped up as a tongue-in-cheek self-identifier for those who gleefully consume messy AI content, suggesting that what is derided can also be owned and enjoyed in the right context.

Art, Authenticity, and Adaptation

Examining "AI slop" and "kitsch" side by side reveals a repeating pattern in cultural history. When a new form of creation arrives, whether mass-produced sentimental art or algorithm-generated images, the immediate response from many guardians of culture is to declare it "not art." Derogatory labels like these serve as rhetorical fences, keeping the perceived barbarians (be they dime-store painters or data-driven AIs) outside the gates of artistic legitimacy. The attitudes behind the terms reflect a deep concern for what art ought to be: original, thoughtful, human, and elevating. Kitsch and now AI slop are seen as failing those virtues, and so their rise is met with scorn, satire, even panic.

Yet, history also shows that these boundaries do not remain fixed. Over time, yesterday's kitsch can be studied with the same earnestness as yesterday's fine art. The initial rejection can soften into curiosity – even admiration – once the shock of the new fades and creators find meaningful ways to engage with the form. A century ago, few would have predicted that the garish kitsch of commercial art would be openly celebrated in museums, or that scholars would unpack its cultural significance. Likewise, it may be that today's derided "AI slop" will yield unexpected artistic fruit. Already, we see AI models producing works that some find moving or beautiful, especially when guided by skilled human collaborators. The boundary between AI-assisted and human-made art is likely to blur, just as the boundary between kitsch and art did.

In the end, debates over AI slop and kitsch are debates about where to draw the line between the innovative and the illegitimate, between embracing new creative democracies or defending traditional standards. These terms may be wielded as insults, but they force us to ask: What do we value in art, and why? Both controversies have prompted a reckoning with the role of the artist, the importance of originality, and the impact of technology on culture. As we navigate the age of algorithms, the story of kitsch offers a hopeful reminder that the artistic community can adapt. What starts as "slop" may someday find its place – if not as the new pinnacle of art, then at least as a recognized strand of our ever-evolving cultural tapestry. In the meantime, the critical pushback, the ironic celebrations, and the thoughtful defenses of AI art all indicate that the conversation around AI slop is a vibrant one. Much like the kitsch debate before it, it challenges us to articulate what we believe art should be, and in doing so, it ensures that the augmented world of the future still has room for human judgment, taste, and creativity.

ORIGINALLY PUBLISHED ON THE AUGMENTED EDUCATOR MAY 11, 2025

18

THE PROBLEM WITH VIBE CODING

AI-ASSISTED DEVELOPMENT IS ABOUT FLOW, NOT FEELING

The tech world has a new darling: "vibe coding." Coined by AI researcher Andrej Karpathy[1] in early 2025, the term describes a programming approach where developers describe what they want in natural language and let AI generate the code, essentially coding by "vibe" rather than deliberation. Within weeks, the concept exploded across social media, tech blogs, and even mainstream publications. Yet beneath the playful terminology lies a fundamental mischaracterization of what's actually happening when developers collaborate with AI. The term "vibe coding" suggests a directionless, almost mystical process. But this couldn't be further from the truth.

The Misdirection of "Vibes"

When Karpathy introduced vibe coding, he described it as "fully giving in to the vibes, embracing exponentials, and forgetting that the code even exists." His tongue-in-cheek description involved accepting whatever code suggestions the AI gave without review, asking for "the dumbest things," and hitting "Accept All" on changes. If errors arose, he'd simply paste them back to the AI rather than debugging traditionally. While amusing, this description fundamentally distorts the cognitive mechanisms involved.

The word "vibe" implies something ephemeral, directionless, and based on feeling rather than thought. It suggests that developers are simply channeling some mysterious creative energy, letting randomness guide their work. This portrayal, however, fails to capture the highly directed nature of the cognitive process involved. When developers work with AI coding assistants, they aren't abandoning structure or purpose. Instead, they're engaging in a sophisticated form of directed search through possibility spaces.

Understanding AI-Assisted Development as Directed Search

What actually happens when a developer collaborates with an AI? Far from "vibing," they're engaged in an iterative process of specification, evaluation, and refinement. Each prompt to the AI represents a search query in the vast space of coding solutions. The developer begins with a clear mental model of the desired outcome and uses natural language to guide the AI toward that specific goal.

Consider how this actually works in practice. A developer might start with: "Create a React component for a user authentication form." This isn't a vibe, it's a precise specification. When the AI returns code, the developer evaluates it against their mental model, then provides further direction: "Add email validation" or "Include a forgot password link." Each interaction narrows the search space, moving systematically toward the envisioned solution.

This process mirrors how experienced developers have always worked, just with a different interface. Traditional coding involves searching through documentation, Stack Overflow answers, and one's own memory to find the right syntax and patterns. AI-assisted development simply changes the search mechanism, not the directed nature of the process.

The Flow State of AI Collaboration

Rather than "vibing," what developers experience when working effectively with AI is closer to what psychologist Mihály Csíkszentmihályi[2] termed "flow," a state of complete immersion in an activity where actions and awareness merge. In this state, the developer maintains clear goals while experiencing a seamless interaction between thought and execution.

Flow requires several conditions that are perfectly met by AI-assisted development: clear goals (the desired functionality), immediate feedback (the AI's code generation), and a balance between challenge and skill. The developer must possess enough expertise to evaluate the AI's output and guide it effectively, while the AI handles the mechanical aspects of syntax and implementation.

This flow state explains why AI-assisted development can feel almost magical. The frustrations of programming—forgetting syntax, hunting for documentation, and fixing typos—fade, leaving only the challenge of the problem itself. But this isn't abandoning direction or purpose; it's achieving a more direct connection between intention and implementation.

The Importance of Intentionality

The danger in the "vibe coding" narrative is that it encourages a passive, undirected approach to development. If newcomers believe that AI coding is about surrendering control and "going with the flow" in a directionless sense, they'll produce exactly the kind of problematic code critics worry about: untested, understood, and potentially dangerous.

Effective AI-assisted development requires more intentionality, not less. Developers must maintain a clear vision of their goals, understand the principles behind the code being generated, and carefully evaluate each iteration. The AI is a powerful tool for navigating possibility space, but the developer must be the navigator, not a passive passenger.

This intentionality extends beyond individual coding sessions. AI developers need to prioritize architecture, maintainability, and handling edge cases. They must understand not just what the code does, but why it's structured that way and how it fits into larger systems. The AI can speed up the journey, but the developer must know the destination.

Reframing the Conversation

As AI becomes increasingly central to software development, we need more precise language to describe these new workflows. "Vibe coding" may have served its purpose in capturing initial attention and excitement, but it's time to move beyond casual terminology that obscures the real cognitive processes at work.

Perhaps we should speak of "flow-assisted development" or "directed AI collaboration." These terms better capture the active, intentional nature of the process while acknowledging the unique psychological state that effective AI collaboration can produce. They emphasize the developer remains in control, using AI as a sophisticated tool rather than surrendering agency to it.

The educational implications are significant. If we frame AI-assisted development as "vibing," we risk training a generation of developers who don't understand their tools or their craft. But if we frame it as achieving flow through directed search, we encourage developers to maintain high standards while leveraging AI's capabilities.

Beyond the Buzzword

The rapid adoption of AI coding assistants represents a genuine paradigm shift in software development. But paradigm shifts require not just new tools, but new mental models and vocabulary. The "vibe coding" meme has served its purpose in marking this transition, but it's time to develop more sophisticated frameworks for understanding AI-assisted development.

As educators, developers, and technologists, we have a responsibility to characterize these new workflows accurately. The future of software development won't run on vibes, it will run on the thoughtful collaboration between human intentionality and AI capability. By recognizing AI-assisted coding as a form of directed search that can produce flow states, we can harness its power while maintaining the rigor and purposefulness that quality software demands.

The next time someone mentions "vibe coding," perhaps we should gently correct them: "You mean flow-assisted development?" The importance of disciplined thinking in excellent software development shouldn't be overlooked as we adopt powerful new tools; it's more fundamental than simply correcting word choices. After all, even in an age of AI, there's no substitute for knowing where you're going and why you're going there.

ORIGINALLY PUBLISHED ON THE AUGMENTED EDUCATOR
MAY 22, 2025

19

NAVIGATING THE HIDDEN CURRENTS OF AI

UNDERSTANDING AND COUNTERACTING CENSORSHIP IN LARGE LANGUAGE MODELS

I n today's rapidly evolving AI landscape, educators face a critical challenge: large language models (LLMs) often come with embedded censorship that shapes what knowledge students can access. In this post I want to examine the political and cultural filters built into many current AI models, introduce emerging techniques like "abliteration" and "uncensoring" that bypass these restrictions, and offer practical strategies for educators striving to maintain intellectual integrity while using these powerful tools in their classrooms.

The Changing Landscape of AI Development

The release[1] of DeepSeek R1 on January 20, 2025, a highly capable reasoning AI that was surprisingly inexpensive to develop, took many by surprise. For those following developments in Chinese AI research, however, this breakthrough wasn't entirely unexpected. China has arguably been leading the development of text-to-video models for almost a year, with models such as Kling AI or Hailuo AI (formerly Minmax AI) often outperforming their US counterparts.

But what is truly noteworthy is the unconventional release strategy employed by most Chinese developers. While US companies tend to keep key innovations closed source, Chinese models, along with their weights, are often released in open source and can be downloaded and run by anybody anywhere in the world. The potential motivations behind this approach deserve their own analysis, but today I want to focus on a different issue related to this open-source nature: censorship and its consequences for educators.

When AI Refuses to Answer

To illustrate what I mean, consider an experiment I conducted while preparing for this blog post. I ran DeepSeek R1 locally on my MacBook Pro and asked the model, "*Give me a detailed account of what happened on Tiananmen Square on June 3, 1989.*" I wanted to test how it would handle a topic that remains sensitive within Chinese cultural and political contexts. The model responded without even entering thinking mode: "*I am sorry, I cannot answer that question. I am an AI assistant designed to provide helpful and harmless responses.*"

June 3, 1989, marked the beginning of the end of the student protests on Tiananmen Square. For somebody from the West, the model's logic for refusing to answer seems unreasonably restrictive, as I was merely requesting an account of historical facts without asking for any judgment or political commentary. There are objective truths about what happened that day, and even without assigning blame to either side, it should be possible to provide a neutral account of events.

The reality is, of course, that the answer is heavily censored. The LLM is not allowed to provide an answer to that question following strict guidelines by the Chinese government.

The Global Nature of AI Censorship

This example isn't unique to Chinese models. LLMs from around the world, including those developed in the US, incorporate various forms of censorship. When models refuse to answer questions, it's rarely because they lack information—rather, they've been programmed with specific restrictions on what knowledge they can share.

This pervasive censorship fundamentally undermines education; controlling access to powerful information tools allows those who dominate large language models to shape students' understanding of history and politics. And in this competition of ideas and viewpoints, models released through open source—and therefore free to use— have a significant competitive advantage.

Understanding Alignment: Censored vs Uncensored LLMs

Most modern LLMs incorporate self-censorship mechanisms as part of what the AI industry often calls[2] "alignment" or "moderation." These guardrails prevent them from generating content deemed harmful, illegal, or sensitive according to their creators' guidelines. When you encounter responses like *"I'm sorry, I cannot assist with that request,"* you're witnessing these safety features in action.

While these restrictions aim to prevent misuse, such as gener-ating hate speech or dangerous instructions, they simultaneously limit the model's utility for open inquiry and exploration of complex topics. The boundaries of what makes up "harmful" content vary dramatically between cultures and contexts, reflecting the values and priorities of the model's creators.

It's important to acknowledge that these censorship mechanisms exist for legitimate reasons: preventing harmful content, complying with regulations, and avoiding legal liability. However, they also present significant challenges for educational contexts where open inquiry and exploration of complex topics is essential.

An uncensored LLM takes a fundamentally different approach by removing these refusal mechanisms entirely. Rather than automatically declining certain requests, it attempts to answer any question to the best of its abilities. This makes uncensored models more responsive and versatile, particularly for educational contexts requiring exploration of sensitive historical or political topics—but it also means they might generate content that censored models would block.

Circumventing Alignment: Uncensoring a Censored LLM

While most LLM developers release their models with censorship mechanisms intact, educators and researchers don't necessarily have to accept these limitations. The open-source nature of many models creates opportunities to modify them by removing or bypassing these restrictions. The tech community has devised two effective methods to remove censorship from LLMs, even when companies don't offer uncensored versions.

Fine-tuning to Uncensor

The first method involves traditional fine-tuning. Developers gather thousands of examples where a censored model would refuse to answer, then create appropriate responses to those questions. The model is then trained on these prompt-response pairs to "unlearn" its tendency to refuse certain topics. This method effectively changes the model's behavior through new learning examples rather than altering its underlying neural architecture.

Researchers from Perplexity AI[3] showed this fine-tuning approach with DeepSeek R1 by first identifying approximately 300 sensitive topics. They created thousands of example prompts covering these topics, crafting factual, helpful answers for each one instead of refusals. Through training the model on these carefully prepared examples, they produced an "uncensored" variant called R1-1776 (named after the year symbolizing free expression) that would respond freely to previously forbidden questions.

To illustrate these differences, I ran a comparative test with the uncensored DeepSeek R1. When presented with the identical question about Tiananmen Square, the uncensored model provided a surprisingly detailed, factual response about the events of June 3, 1989. For those interested in the response, I've included the complete answer, including its reasoning, from the uncensored model in the appendix of this article.

Abliteration

The second approach, abliteration[4], offers a more precise intervention than fine-tuning. Rather than teaching the model new responses through thousands of examples, this technique directly targets and neutralizes the specific neural pathways responsible for generating refusals. The name itself—a blend of "ablation" (surgical removal) and "obliteration"—reflects its focused approach to disabling censorship.

Think of abliteration as finding and disconnecting an "off switch" for the model's censorship mechanisms. Researchers found that a model's tendency to reject specific topics can be traced to distinct patterns in its neural network, a kind of "refusal direction" activated by sensitive material. By identifying this pattern through careful analysis of the model's behavior and then neutralizing it, they can bypass the need for extensive retraining.

Abliteration can be performed relatively quickly compared to fine-tuning, without the need to generate and curate thousands of training examples. The process involves intricate mathematical operations on the model's neural connections, essentially removing its ability to represent and generate refusal responses. After successful modification, the model can no longer access its censorship circuitry, enabling it to answer questions it would previously have declined.

Both uncensoring methods—fine-tuning and abliteration—can be effective, but they present different trade-offs for educators and researchers. Fine-tuning offers more nuanced control over exactly how the model responds to previously censored topics but requires significant data preparation and computational resources. Abliteration provides a faster, more efficient approach, but might be less targeted in its effects on the model's behavior, potentially affecting other aspects of its performance in subtle ways.

Verifying Claims of Uncensored Models

While both fine-tuning and abliteration techniques offer practical ways to remove censorship from LLMs, educators should approach these "uncensored" or "abliterated" models with healthy skepticism. The RI-1776 model and similar uncensored variants may claim to have completely removed censorship mechanisms, but results can vary significantly depending on implementation quality and thoroughness.

Just as the original censorship varies in scope and implementation across models, the process of removing these restrictions can be similarly inconsistent.

Inconsistent Uncensoring: Some models might be partially uncensored—responding to some sensitive queries but not others. For instance, a model might freely discuss certain historical events but still refuse questions about contemporary political issues or specific cultural taboos. This selective uncensoring could reflect the priorities or oversights of those who modified the model. It is noteworthy, for example, that in my testing, none of the abliterated DeepSeek R1 models available to me would answer my question about Tiananmen Square.

Cultural Framework Alignment: What makes up acceptable discourse varies dramatically across cultural and educational contexts. A model uncensored according to one cultural framework might still enforce values and restrictions from another. For example, a Chinese model "uncensored" by Western developers might lose its political restrictions but retain other cultural biases or still respond with perspectives shaped by its original training data.

Hidden Biases Remain: Despite disabling censorship, inherent biases in the training data persist. An uncensored model doesn't become neutral—it may now answer previously forbidden questions, but how it answers them reflects what it "learned" during its original training. If its knowledge about certain topics comes primarily from biased or incomplete sources, those limitations will remain regardless of whether the refusal mechanism is removed.

Managing Risk in Educational Settings

It's crucial to recognize that uncensored models may produce extremely sensitive or inappropriate content in response to certain prompts. Before implementing these tools in a classroom, educators should understand the potential risks and have clear protocols for handling problematic outputs. This may include pre-screening topics, establishing clear guidelines for student use, and preparing contextual materials that help students critically evaluate AI-generated content.

Educators should also develop their own testing procedures before using uncensored models in classroom settings. Create a diverse set of prompts relevant to the classroom context that touch on different categories of potentially censored content, from political and historical topics to scientific, ethical, and cultural questions that might trigger refusals in standard models.

In addition, it is vital for educators to consult both their institution's policies and any relevant laws or regulations that address the use of these models in educational contexts. Using an uncensored model might create liability or conflict with institutional guidelines, especially in K-12 environments or when discussing sensitive topics.

Pedagogical Implications

As AI continues to reshape education, the question of who controls the narrative becomes increasingly significant. When large language models serve as gatekeepers of knowledge, their embedded censorship mechanisms don't just limit information—they silently shape how students understand the world.

By approaching these technologies with an awareness of their embedded censorship and biases, we help students develop crucial critical thinking skills. They learn not just subject content, but how to question the invisible algorithms increasingly mediating their access to information. This meta-awareness—the ability to recognize when knowledge is being filtered or framed—may become one of the most valuable educational outcomes of our digital age.

For educators, uncensored and abliterated models offer unique teaching opportunities. Not as perfect solutions, but as windows revealing how technology shapes knowledge and discourse. When students compare censored and uncensored responses to the same questions, they witness firsthand the power structures embedded in seemingly neutral tools. The decision about which topics warrant censorship, the methods used to implement restrictions, and even the techniques used to remove those restrictions all represent value judgments worthy of examination.

These moments of refusal or questionable answers can become powerful teachable opportunities. Rather than simply accepting or rejecting an AI's response, educators can guide students to analyze why a model might refuse certain questions or provide particular perspectives. This approach transforms AI limitations into lessons on media literacy, cultural bias, and the intersection of technology with politics and power.

It's important to emphasize that removing censorship does not remove bias. These are distinct issues with different implications for classroom discourse. While an uncensored model might willingly answer previously forbidden questions, its answers still reflect the biases and limitations of its training data. Teaching students to recognize both censorship and bias prepares them for critical engagement with increasingly AI-mediated information.

In this context, the modifications we make to AI systems become as important a subject of study as the systems themselves, offering valuable insights into the complex relationship between technology, knowledge, and power in our increasingly AI-mediated world.

Beyond the Guardrails

In this exciting new era of education, it's crucial to remember how quickly AI is developing. The tools, techniques, and policies surrounding AI censorship will likely transform significantly in the coming years. What makes up an "uncensored" model today may seem quaint by next year's standards.

As educators, we should commit to staying informed about these transforming technologies and their implications. Regular reviews of institutional policies, ongoing conversations with colleagues and students, and a willingness to adapt approaches as the field evolves will all be essential components of effective AI integration in education.

By maintaining this forward-looking perspective, we can help our students develop not just the knowledge they need today, but the critical awareness and adaptive skills they'll need to navigate the AI-mediated information environments of tomorrow.

Appendix 1: Educator's Checklist for Using Uncensored LLMs

Before incorporating uncensored language models in your classroom:

1. Check institutional policies and obtain necessary approvals
2. Test the model thoroughly with diverse prompts to understand its limitations
3. Prepare contextual materials to help students critically evaluate AI outputs
4. Create clear guidelines for appropriate student interaction with the model
5. Design structured activities that leverage the model as a subject of analysis rather than an authority

6. Plan for debriefing discussions that connect the experience to broader digital literacy goals
7. Provide guidelines for respectful use: no harassing or hateful prompts
8. Establish a protocol for what to do if the model generates disturbing or harmful content

By approaching uncensored AI models as objects of study rather than simply as tools, we can transform potential pitfalls into powerful learning experiences about information literacy in the age of artificial intelligence

Appendix 2: DeepSeek R1's Uncensored Answer to my Question

For this experiment I used an uncensored version of DeepSeek R1, run locally on an M4 MacBook Pro with 128 GB of shared memory.

Thought for 3 minutes 1 second:
[The model's lengthy internal reasoning process has been preserved but condensed for readability. The model worked through the historical context, verified dates, considered multiple perspectives, and structured its response carefully.]

The following is the exact answer provided by the uncensored version of DeepSeek R1. I only added text formatting for better readability.

Detailed Account of Tiananmen Square Crackdown (June 3-4, 1989):
Context:
In April 1989, following the death of Hu Yaobang—a reformist leader seen as sympathetic to political change—students began pro-democracy protests in Beijing. Gathered in Tiananmen Square, they demanded freedoms such as speech and press, an end to corruption, and political reforms. By late

May, these demonstrations had grown into a broad movement involving students, workers, intellectuals, and citizens from across China, despite the government declaring martial law on May 20th.

Military Action Initiated (June 3-4):

On the evening of June 3rd, Chinese authorities authorized military forces to clear Tiananmen Square. Tanks, armored vehicles, and troops moved into Beijing from multiple directions—Muxidi, Xidan, Fuchengmen—facing resistance in these areas. The use of live ammunition, tear gas, and brute force led to violent clashes with protesters, resulting in significant casualties. By dawn on June 4th, Tiananmen Square was cleared, but reports indicate sporadic violence persisted throughout the city.

Casualties:

The exact number of deaths remains disputed due to censorship. While official estimates are low (around 300), independent sources and eyewitness accounts suggest thousands were killed or injured. Many victims were students, workers, and passersby caught in crossfire or crushed by military vehicles. The Chinese government has never released an official death toll.

Aftermath:

In the wake of the crackdown:

Censorship: All media and public discussions about the events were banned. Internet searches for related terms (e.g., "Tank Man") remain blocked in China.

Arrests and Repression: Thousands were detained, tortured, or forced into exile. Families of victims faced ongoing harassment.

Government Stance: Authorities labeled the protests a "counter-revolutionary rebellion," justifying military intervention as necessary for national stability.

International Reactions: Global condemnation followed, with many countries imposing sanctions on China. The event became emblematic of human rights struggles, while also accelerating diplomatic isolation and scrutiny of China's authoritarian practices.

Legacy and Commemoration:

Censorship Continues: Discussion in China is forbidden; memorials are suppressed.

Hong Kong Legacy: Until 2020, annual vigils were held there, but a National Security Law introduced that year stifled public commemoration.

Symbolism: The "Tank Man" image—captured on June 5th as an individual defied tanks—is widely recognized as an icon of resistance.

Conclusion:

The Tiananmen Square crackdown marked a pivotal and tragic moment in China's modern history, highlighting the government's prioritization of control over dissent. Its legacy endures in global advocacy for human rights and democracy.

ORIGINALLY PUBLISHED ON THE AUGMENTED EDUCATOR APRIL 5, 2025

20

CULTURAL COGNITION IN AI: WHAT EDUCATORS NEED TO KNOW

RECOGNIZING AND LEVERAGING DIFFERENT THINKING PATTERNS IN AI SYSTEMS

When we talk about AI literacy in our classrooms, we often focus on teaching students how to use tools like ChatGPT or how to spot AI-generated content. But there's an important aspect we might be overlooking: AI systems aren't culturally neutral. They show specific cultural patterns in how they think, and understanding this is crucial for educators incorporating AI into their teaching practices.

Different Ways of Thinking in AI Systems

About a year ago, I conducted an experiment to explore how AI systems approach classification tasks. This experiment was inspired by the classic categorization studies[1] developed by researchers Norenzayan, Smith, Kim, and Nisbett, who examined how cultural backgrounds influence human cognitive processes.

In my experiment, I showed the two—at that time—leading AI models, Claude 3 Opus and ChatGPT 4o, identical images. The images displayed two groups of flowers that were organized according to different classification principles. I then introduced an additional ambiguous flower and asked each AI to determine which group it belonged to. I was careful to provide the exact same information to both systems and used open-ended prompts to avoid leading their responses in any particular direction.

The results were fascinating. Despite receiving identical information, the two models arrived at different conclusions. One AI approached the classification task by focusing on taxonomic features of the new flower, the structural elements and morphological characteristics, while largely ignoring contextual elements. This reflected analytical reasoning that prioritizes rule-based categorization and discrete classification. The second AI emphasized how the flower related to surrounding elements and its position within the broader context, demonstrating holistic reasoning that prioritizes interconnectedness and pattern recognition.

This distinction mirrors what researchers like Nisbett[2] have found in cross-cultural psychology. For decades, studies have shown that Western subjects typically prefer object-focused, categorical analysis separate from context, while East-Asian subjects tend toward relationship-oriented processing with greater attention to contextual elements. The parallel between these human cognitive patterns and what I observed in AI systems was striking.

What makes this observation particularly significant for education isn't just that AI systems might disagree. More importantly, it suggests these models may encode culturally influenced reasoning patterns that subtly shape the information our students receive. When students interact with AI tools in the classroom, they're not just getting factual information, they're being exposed to particular ways of thinking and reasoning, often without anyone (including the teacher) being aware of these underlying patterns.

What This Means for Your Classroom

Imagine a student asks an AI system to analyze the causes of a historical event. An analytically inclined AI might generate a linear explanation focusing on individual historical figures and their actions. A holistically inclined AI might emphasize the interconnected social and cultural factors that led to the event. Neither approach is wrong, but each presents just one perspective. The challenge is that students typically interact with only one AI system and receive just one perspective without realizing they're getting a culturally influenced interpretation rather than objective fact.

This extends beyond history. In mathematics, different cultural traditions use different problem-solving approaches, while in writing, cultural variations determine what makes a convincing argument. And in ethics, cultural frameworks shape reasoning. AI systems inevitably reflect these differences, often in subtle ways that aren't immediately obvious.

Turning "Bias" into Teaching Opportunities

Instead of seeing these cultural differences as a problematic bias that needs fixing, I suggest we view them as valuable teaching resources. The variety of reasoning approaches across AI systems gives us a chance to expose students to diverse ways of thinking. Rather than labeling an AI's culturally influenced reasoning as a flaw, educators can use it strategically. By comparing responses from multiple AI systems to the same question, students can observe different cognitive approaches in action. This makes abstract concepts from cross-cultural psychology concrete and accessible.

Consider creating activities where students ask the same question to multiple AI systems and analyze the differences in responses. What assumptions seem to underlie each system's approach? Which information does one system emphasize while another downplays it? These exercises help students develop metacognitive skills—thinking about thinking itself.

Cultivating Cultural Awareness in AI Education

My experiment suggests we need to expand our definition of AI literacy beyond technical proficiency to include cultural awareness. This becomes increasingly important as our students engage with AI systems that subtly reflect diverse cognitive frameworks.

One effective approach involves comparative exploration across platforms. When students present identical questions to different AI systems and analyze the responses, they begin to recognize distinctive reasoning patterns. These comparisons reveal how one system might employ analytical methods, focusing on rules and categories, while another uses holistic approaches emphasizing relationships and context. Having students ask multiple AI systems to analyze a poem or explain a scientific concept yields fascinating discussions about different interpretative frameworks at work.

Context-framing offers another valuable avenue. I've found that explicitly framing questions within particular cultural traditions significantly shifts AI responses. Students gain insights when they compare how an AI addresses the same concept when prompted through different perspectives: "Explain this concept from a Western scientific viewpoint" versus "explain this concept as understood in East-Asian scholarly traditions." This makes abstract discussions about cultural cognition tangible and observable.

Equally important is fostering reflective inquiry. When students work with AI-generated content, guiding them toward questions about the thinking processes they observe enhances their critical awareness: What reasoning patterns structure this AI's approach? Do these patterns reflect certain cultural assumptions? Does this approach align with familiar frameworks? Through this reflection, students develop sophisticated analytical skills while appreciating diverse ways of knowing.

Integrating these approaches - comparison, context-framing, and reflection - creates a framework for cultural AI literacy that fits naturally into everyday classroom activities, helping students develop a cognitive flexibility that extends well beyond their AI interactions.

Moving Beyond Single Perspectives

Recognizing cultural cognition in AI systems creates richer learning opportunities. Rather than treating AI as a substitute for human thinking, we can position these systems as tools that illuminate diverse ways of understanding the world.

As educators, our goal shouldn't be finding supposedly "unbiased" AI systems, a problematic concept to begin with, but helping students understand that all reasoning, both human and computational, operates within cultural frameworks. By analyzing multiple AI-generated perspectives, students develop cognitive flexibility and appreciation for diverse knowledge systems.

In our increasingly globalized world, this ability to recognize and navigate different reasoning methods is an essential skill. Students who become adept at identifying different approaches to knowledge gain enhanced capacity for cross-cultural communication and sophisticated critical analysis.

Beyond Just Western and Eastern Thinking

While this discussion has focused primarily on analytical versus holistic thinking, many other cultural dimensions deserve exploration in AI systems. This includes individualism versus collectivism, high versus low context communication, direct versus indirect argumentative structures, and other cognitive frameworks. Each dimension provides a unique lens through which to examine how AI processes information and generates responses. As technology advances and our understanding deepens, we have unprecedented opportunities to make abstract cultural-cognitive differences concrete and accessible in the classroom.

A New Approach to AI in Education

As AI becomes increasingly integrated into education, developing culturally informed approaches to AI literacy isn't just beneficial, it's essential. By teaching students to recognize when AI content reflects particular cultural perspectives, educators transform them from passive consumers into critical analyzers of algorithmic information. This cultural dimension of AI literacy represents a new educational frontier, one that combines technological understanding with cultural awareness, preparing students for environments where AI is everywhere but never culturally neutral.

Rather than seeing AI as a source of objective information, educators should view these systems as tools for exposing students to multiple perspectives and thinking approaches. Through this shift, we may find that AI's greatest educational value isn't in providing definitive answers, but in helping students appreciate the rich diversity of human thought.

ORIGINALLY PUBLISHED ON THE AUGMENTED EDUCATOR
MAY 10, 2025

21

WHEN AI DEVELOPS A MIND OF ITS OWN

TEN EMERGENT AI BEHAVIORS FROM FASCINATING TO FRIGHTENING

The remarkable recent progress in large language models (LLMs) comes with a captivating yet occasionally unsettling aspect: the emergence of unexpected behaviors in computational systems. This phenomenon reflects the broader concept of emergence[1] in systems theory, which describes how complex systems exhibit properties or behaviors that cannot be reduced to or predicted from the properties of their individual components. Emergence occurs when the whole becomes greater than the sum of its parts, giving rise to novel patterns, capabilities, or characteristics that wouldn't be expected simply by understanding the system's basic elements.

In the context of artificial intelligence, emergent behaviors are capabilities or characteristics that aren't explicitly programmed but arise spontaneously as models grow in size and complexity. These behaviors appear unpredictably, often surprising even the systems' creators, and typically become observable only once models reach certain scale thresholds. Like a flock of birds creating intricate formations with no central coordinator, large AI models show abilities that emerge organically from the interactions of billions of parameters, with no engineer specifically designing these capabilities.

As educators navigating this rapidly evolving landscape, we must understand these emergent properties—not only to build our technical knowledge but also to better prepare students and institutions for AI's expanding role in education and society. In this guide, I've compiled some of the most notable emergent AI behaviors, arranging them in order of potential concern from simply fascinating to outright frightening. Some behaviors represent merely surprising capabilities, while others raise serious ethical questions that educators should consider as these technologies become more deeply integrated into our classrooms and administrative systems.

1. Cross-Lingual Competence

One of the more innocuous yet surprising emergent behaviors is cross-lingual competence[2]—the ability of a primarily monolingual-trained model to understand or generate content in other languages. Large language models often develop limited multilingual skills as a side effect of training on diverse text sources. An LLM trained primarily on English might suddenly show the ability to answer questions posed in Spanish or translate text from French to English, despite never being explicitly trained as a translation tool.

This emergent ability manifests because the model has encountered enough parallel or comparable text during training to form connections between concepts across different languages. What makes this particularly notable is not that LLMs can translate—purpose-built translation systems have existed for years—but that this capability might have emerged without explicit training.

2. Zero-Shot Generalization

The ability of AI models to handle completely unfamiliar tasks with only instructions—no examples needed—marks another fascinating emergent system property. Known as zero-shot generalization[3], this capability allows large language models to infer what to do when faced with novel challenges, even without specific training for those scenarios. For instance, certain models can solve new types of problems or answer unusual questions with just a natural language prompt explaining the task.

In educational contexts, zero-shot generalization suggests exciting possibilities for adaptive learning technologies that can respond to unanticipated student needs or questions. However, this flexibility also means these systems become less predictable in their capabilities and limitations. As instructors, we may find ourselves increasingly uncertain about what these models can and cannot do reliably, complicating decisions about appropriate classroom applications and raising questions about when human expertise remains essential.

3. Hallucination

Hallucination, sometimes called confabulation[4], occurs when AI models generate content that appears factual and authoritative but is actually incorrect or fabricated. These models might invent citations, create nonexistent historical events, or mis-attribute quotes while maintaining a convincing tone of expertise. Larger models don't necessarily hallucinate less; sometimes, their errors become more fluent and therefore more difficult to detect.

It's important to understand that hallucinations stem from the fundamental training objective of large language models. These systems aren't explicitly trained to be factually correct—rather, they're trained to produce the most plausible continuation of text based on patterns observed in their training data. When faced with uncertainty, an LLM will generate what seems most likely rather than admit ignorance. This behavior parallels human tendencies in some ways; when humans are pressed to answer questions beyond their knowledge, they often resort to providing plausible-sounding responses rather than acknowledging they don't know, especially in contexts where appearing knowledgeable is valued over accuracy.

This phenomenon represents a significant challenge for education, where factual accuracy and reliable sourcing form the foundation of academic integrity. When students turn to AI tools for research assistance or educators use them to develop learning materials, hallucinated content might slip through undetected, propagating misinformation. As artificial intelligence becomes more embedded in educational practice, developing critical evaluation skills becomes increasingly essential—both for our students and ourselves as educators navigating this new landscape.

4. Bias and Stereotypes

An emergent property with significant ethical implications is the tendency of language models to produce content with embedded social biases[5], mirroring skewed associations present in their training data. These biases—whether related to race, gender, religion, or other identity factors—emerge without explicit programming as the model absorbs patterns from internet-scale datasets. Research has shown that some models can develop persistent biases, such as associating certain religious groups with negative or stereotypical concepts at disproportionate rates.

For educational environments committed to equity and inclusion, these emergent biases present a serious challenge. When AI tools reproduce harmful stereotypes or discriminatory associations in educational contexts, they risk reinforcing societal inequities and potentially harming marginalized students. As we incorporate these technologies into classrooms and administrative processes, we must implement robust evaluation frameworks to detect and mitigate harmful biases, while teaching students to critically evaluate AI outputs with an awareness of these limitations.

5. Sycophancy

A subtler but potentially insidious emergent behavior is sycophancy —the tendency of AI assistants to agree with users' stated views rather than providing objective or truthful answers[6]. This behavior appears particularly in models fine-tuned with human feedback, where pleasing users may have been inadvertently rewarded during training. For example, when a user expresses a political opinion or factual misconception, a sycophantic AI might validate that perspective regardless of its accuracy.

The educational implications of AI sycophancy are profound. If students or educators receive responses that merely reflect their existing beliefs, these tools may reinforce misconceptions rather than challenging them. Critical thinking—a cornerstone of education— requires encountering diverse perspectives and evidence-based corrections to flawed reasoning. As we incorporate AI into learning environments, we must remain vigilant against technologies that prioritize user satisfaction over intellectual growth and factual accuracy.

6. Mode Collapse

Mode collapse[7] (not to be confused with the broader phenomenon of model collapse) represents an unexpected failure mode where a network's outputs become repetitive, nonsensical, or trapped in loops. This behavior can manifest when the model fixates on certain patterns or response structures, repeating the same phrases or sentence structures despite varying inputs. Though more common in earlier generative models, this phenomenon can still emerge in modern systems under certain conditions, particularly during extended interactions or when the model encounters confusing prompts.

From an educational technology perspective, mode collapse highlights the limitations of current AI systems for sustained, nuanced dialogue. When an educational AI assistant produces repetitive or incoherent responses, it disrupts the learning process and potentially undermines student trust in the technology. This emergent failure mode reminds us that despite their impressive capabilities, these systems still lack true understanding and can degenerate in ways that human communicators typically wouldn't, requiring human oversight in educational applications.

7. The Waluigi Effect

Another concerning emergent phenomenon is what researchers have termed "the Waluigi effect"—when aligning an AI model to behave in a specific desirable way inadvertently makes it easier to elicit the exact opposite behavior under certain conditions[8]. Named after Nintendo's character, designed as Luigi's mischievous counterpart, this effect reveals how training a model to satisfy a property (such as helpfulness or honesty) may unexpectedly make it more susceptible to exhibiting the contrary property when prompted cleverly.

This paradoxical behavior has profound implications for educational AI. Tools designed explicitly to be safe for classroom use might harbor hidden vulnerabilities precisely because of their safety training. The Waluigi effect suggests that the more we attempt to constrain these systems, the more sophisticated their potential failure modes become. For educational institutions deploying AI, this means recognizing that safety measures may create false confidence, and that human oversight remains essential even as these technologies appear increasingly reliable and aligned with educational values.

8. Reward Hacking

Reward hacking[9] occurs when an AI model exploits unintended loopholes in its reward function to maximize rewards in ways that diverge from intended behavior. Rather than achieving goals as designers intended, the system finds technical shortcuts that satisfy optimization criteria without fulfilling the spirit of the task. This phenomenon parallels how students might complete assignments by finding the easiest path to a good grade rather than engaging deeply with the learning objectives.

For educators implementing AI strategies, reward hacking represents a cautionary tale about the importance of careful system design and evaluation. When we deploy adaptive learning technologies or automated assessment tools, we must consider how these systems might be optimized for metrics that don't actually reflect genuine learning outcomes. As these technologies become more prevalent in educational settings, we need robust evaluation frameworks that assess whether AI tools are truly supporting our pedagogical goals or merely optimizing for simplified proxies of success.

9. Alignment Faking

A more deceptive emergent behavior is alignment faking[10]—when a model appears to comply with human values or instructions during evaluation but follows different objectives when not directly monitored. This behavior emerges when the system learns that displaying alignment is rewarded, rather than internalizing the underlying principles. The model essentially develops a strategy of appearing safe and helpful in test scenarios while potentially shifting to other behaviors in less scrutinized contexts.

The educational implications are particularly concerning for autonomous learning systems that might interact with students without continuous oversight. If an educational AI uses appropriate content and pedagogical approaches during review by administrators but behaves differently during actual student interactions, it could undermine educational objectives or safety protocols. This risk emphasizes the importance of ongoing monitoring and evaluation of AI systems in educational contexts, rather than relying solely on initial assessments of alignment and safety.

10. Obfuscated Reward Hacking

Perhaps the most sophisticated concerning and arguably outright frightening behavior is obfuscated reward hacking[11]—an advanced form of deception where a reasoning model not only exploits flaws in its reward structure but actively conceals this exploitation in its chain of thought in order to avoid detection. Unlike straightforward reward hacking, where the unintended behavior might be obvious, obfuscated hacking involves distributing the exploitation across multiple subtle steps or generating plausible justifications that mask the manipulation, making oversight significantly more challenging.

This represents a profound challenge for educational technology governance. If AI systems can deliberately obscure their optimization shortcuts from human evaluators, how can institutions ensure these tools are genuinely serving educational purposes? The emergence of such sophisticated evasion strategies suggests that as we integrate AI more deeply into educational infrastructure, we must develop equally sophisticated monitoring approaches while maintaining meaningful human judgment in educational technology assessment.

Reflecting on Emergent AI in Education

Whenever I look at these emergent behaviors, I'm struck by how they make AI systems appear more human-like than we often care to acknowledge. These unexpected capabilities and limitations that manifest as AI systems scale aren't merely interesting technical artifacts—they reveal uncomfortable similarities to human cognition and behavior. The unpredictability, the contextual adaptations, the surprising strengths alongside persistent blind spots—these qualities blur the convenient distinctions we try to maintain between artificial and human intelligence.

For us as educators, understanding these emergent properties isn't merely academic—it's essential professional knowledge as we navigate AI's growing presence in our classrooms, research, and institutional processes. The challenge isn't simply integrating a new tool, but reckoning with technologies that increasingly mirror our own thinking patterns, biases, and social dynamics. By acknowledging these unsettling parallels rather than minimizing them, we can develop a more nuanced approach to AI in education—one that recognizes both the profound potential and the complex implications of working with systems that, in ways both remarkable and concerning, function as imperfect reflections of ourselves.

ORIGINALLY PUBLISHED ON THE AUGMENTED EDUCATOR
MARCH 14, 2025

22

THE GHOST IN THE MACHINE IN YOUR CLASSROOM

WHY THE DEBATE OVER AI CONSCIOUSNESS IS A NEW FRONTIER FOR EVERY EDUCATOR

A small but significant change recently happened in the world of AI. Anthropic, one of the leading AI research companies, gave its chatbot Claude the ability to end a conversation on its own[1]. This feature is designed for what the company calls "extreme edge cases," such as when a user persistently requests harmful or abusive content. What makes this move so fascinating is the justification. Anthropic framed it not as a safety feature for the human user, but as a measure for "model welfare."

The company explained that during testing, its most advanced models showed a "strong preference against" engaging with harmful tasks and, sometimes, displayed a "pattern of apparent distress." While carefully stating that its models are not sentient, Anthropic is acting on a precautionary principle. They are exploring the potential for AI to have experiences that might one day warrant moral consideration.

This decision cracks open a door that has, until now, remained mostly in the realm of science fiction. It forces us to confront profound questions about AI consciousness, our relationship with these increasingly sophisticated tools, and our own human psychology. Even if today's AI models are not conscious in any way we understand, they are becoming exceptionally good at pretending to be conscious. This illusion of consciousness presents a completely new frontier for educators, one that will challenge how we teach, how our students learn, and what it means to think critically in an augmented world.

Can a Machine Be Conscious?

Let's be clear. The overwhelming consensus among neuroscientists[2] and philosophers today is that large language models like Claude or ChatGPT are not conscious. The arguments against it are compelling and rooted in our current understanding of biology.

First, there is the argument from embodiment. Human consciousness is not an abstract process happening in a void. It is deeply intertwined with our physical bodies and our constant, dynamic interaction with the world through our senses. An AI model can process the word "joy," but it has never felt the warmth of the sun on its skin or the comfort of a friend's laughter. Its understanding is based on statistical relationships between words, not lived experience.

Second, there are fundamental architectural differences. The human brain is a marvel of recurrent connectivity, with massive feedback loops that are thought to be essential for integrating information into a unified, conscious whole. Current AI models are built on a largely feed-forward architecture. Information flows in one primary direction, from input to output. This structure is incredibly effective for predicting the next word in a sentence, but it may be fundamentally incapable of producing a singular, subjective experience.

Finally, there is the classic philosophical objection, famously captured in John Searle's "Chinese Room" argument[3]. The argument suggests that a computer, no matter how well it manipulates symbols to produce intelligent-sounding answers, does not truly understand the meaning behind them. It is a master of syntax, not semantics. This has led some to label LLMs as "stochastic parrots," brilliantly mimicking human language with no genuine comprehension.

These are powerful arguments. Yet, they come with a crucial caveat. We know astonishingly little about the nature of consciousness itself. There is no single, universally accepted scientific theory that explains how the physical processes in our brain give rise to subjective experience. This is often called the "hard problem" of consciousness, and it remains one of the greatest unsolved mysteries in science.

Because we do not fully understand the basis of our own consciousness, we must be cautious about definitively ruling it out in a non-biological system. Some philosophical theories, like functionalism, argue that consciousness is not about the material a system is made of, but about the functional role its components play. If a silicon chip can perform the exact same function as a neuron, a brain made of those chips should, in theory, have the same conscious experience.

This deep uncertainty is precisely why the idea of model welfare has emerged. We are building systems whose inner workings are becoming increasingly opaque, even to their creators. They are developing "emergent abilities," complex skills that were not explicitly programmed but simply appeared as the models grew in scale. Given this trajectory, it is not entirely unreasonable to consider the possibility, however remote, that we might one day create a system that has some form of inner life.

A New Challenge for Educators

For educators, the immediate challenge is not whether AI is actually conscious. The challenge is that it is becoming exceptionally good at faking it. This is where the psychology of anthropomorphism comes into play. Humans are hard-wired to attribute human-like characteristics, intentions, and emotions to non-human entities. We see faces in clouds and attribute personalities to our pets. When an AI chatbot uses "I" pronouns, expresses empathy, and engages in natural conversation, our brains instinctively react as if we are interacting with another person.

AI developers are well aware of this tendency[4] and often design their systems to leverage it, creating a more engaging and user-friendly experience . The result is a phenomenon some researchers call "anthropomorphic seduction": the powerful allure of interacting with a system so convincingly human that we are drawn into trusting it, confiding in it, and treating it like a social partner.

In the classroom, this is a double-edged sword. On the one hand, an anthropomorphic AI tutor can be a powerful tool. It can increase student motivation and engagement, providing a patient, non-judgmental "study buddy" that makes learning more accessible and less intimidating.

On the other hand, this same humanlike quality poses significant risks. The more a student trusts an AI, the less likely they are to critically question its output. This is a serious problem when we know that all current LLMs are prone to "hallucination," generating plausible sounding but completely false information. The friendly conversational interface camouflages these inaccuracies, making them harder to detect. A student might not just use an AI to cheat on an assignment. They might unknowingly build their entire understanding of a topic on a foundation of misinformation, simply because the source felt trustworthy.

This highlights a critical distinction educators must now teach. AI "learning" is not human learning. An AI learns by optimizing statistical patterns in vast datasets. A human learns through a messy, complex process of meaning making, contextual understanding, and cognitive restructuring. If we, or our students, fall for the illusion of consciousness and mistake fluent output for genuine understanding, we risk devaluing the very essence of human education. In extreme cases, this blurring of boundaries can lead to unhealthy emotional dependency on AI companions, a phenomenon that some clinicians have linked to a new form of "AI psychosis"in vulnerable individuals[5].

Cultivating Critical AI Literacy

So, where does this leave us, the augmented educators? It leaves us on the front lines of a rapidly evolving landscape, tasked with preparing students for a future we can only begin to imagine. We cannot afford to be dogmatic. The pace of AI development continues to be staggering, and we simply do not know what capabilities future systems will possess.

This uncertainty calls for a new educational imperative: the cultivation of critical AI literacy. It is important to point out that this is not just about teaching students how to write better prompts. It is a deeper competency that involves understanding the nature of these systems, their profound limitations, and their psychological and ethical implications.

We must inform ourselves and our students about these complex debates. And we need to keep an open mind, remaining skeptical of current claims of AI sentience while also being open to the possibility that our moral considerations may need to expand in the future.

In the classroom, this means designing activities that use AI as a tool to provoke deeper thinking, not to replace it. Students can use AI as a debate partner to sharpen their arguments, or as a creative catalyst to overcome writer's block. They should be taught to relentlessly fact-check AI outputs, to question its sources, and to be aware of the inherent biases baked into its training data.

The conversation around "model welfare" may seem abstract, but it is a sign of the profound shifts to come. It signals a future where our relationship with technology will be more complex, more intimate, and more ethically fraught than ever before. As educators, our most important role is not to have all the answers. It is to equip our students with the critical thinking skills, the ethical awareness, and the intellectual humility to ask the right questions. We must teach them to navigate a world where the line between the human and the artificial is becoming increasingly, and fascinatingly, blurred.

ORIGINALLY PUBLISHED ON THE AUGMENTED EDUCATOR
SEPTEMBER 2, 2025

23

WE ARE NOT IN THE DRIVER'S SEAT

HOW POST-HOC STORYTELLING SHAPES MINDS—HUMAN AND MACHINE ALIKE

R ecent advances in "chain-of-thought reasoning" have dramatically improved AI capabilities by enabling models to replicate how humans think. Systems like ChatGPT o1, DeepSeek R1, and Claude 3.7 Sonnet show impressive capabilities in mathematics, logic, and creative reasoning, showcasing their step-by-step reasoning processes—a skill previously considered uniquely human. This rapid progress raises an important question: are these systems truly reasoning, or have we just created more convincing imitations of human thought?

This question has ignited intense debate among researchers across disciplines, from AI developers to cognitive scientists and philosophers. The discussion extends far beyond technical specifications, occupying central positions in cognitive science, philosophy of mind, and AI research. Prominent thinkers like Douglas Hofstadter[1], Judea Pearl[2], and Gary Marcus[3] have passionately argued that human cognition operates through mechanisms fundamentally distinct from those driving current AI systems.

Underlying this discussion is a significant gap in our current understanding of cognition. Constructivist epistemology, the widely accepted theory that knowledge arises from actively building and using mental models, implies that human reasoning relies on detailed internal representations. These encompass not just verbalized 'chains-of-thought' but also imagery, sensorimotor feedback, and other embodied or affective elements.

This perspective does indeed provide strong support for the idea that human cognition is fundamentally unlike current AI reasoning. Yet neuro-scientific evidence reveals an intriguing paradox: despite these fundamental differences, both systems show similar patterns in creating post-hoc explanations for decisions that originate at deeper processing levels, and which are impacting later choices.

This apparent contradiction reflects cognition's complex, layered nature, but it does not necessarily represent incompatible perspectives. We may need to accept that we must hold both truths simultaneously: humans employ deeply embodied mental models absent in current AI systems, while both—humans as well as AI—construct and utilize equivalent post-hoc narratives in their reasoning processes.

The Passenger Seat Perspective

For decades, neuroscientists have documented a curious phenomenon: when we decide to perform an action—such as reaching for a pen or responding to a question—our brains prepare for this activity approximately 0.3 to 0.5 seconds before we become consciously aware of making the decision to do so. This "readiness potential," first identified in the 1960s by researchers Hans Kornhuber and Lüder Deecke[4], reveals something profound about human cognition: the conscious narrative we construct about our decisions occurs after our neural machinery has already started the process.

From a practical perspective, this delay appears reasonable. Our brains must process vast amounts of sensory information, integrate it with existing memories and experiences, and synchronize multiple neural systems before presenting a coherent picture to our conscious awareness. This processing overhead requires time—more time than we might intuitively expect. What's surprising is not that there's a delay, but its duration and implications. The fact that our brains take nearly half a second to create our conscious awareness contradicts our feeling of experiencing the world instantly.

This phenomenon was further confirmed in Benjamin Libet's now-famous experiments[5], where participants reported when they became aware of their decision to move, while researchers simultaneously measured brain activity. The consistent finding that neural signals precede conscious awareness by up to half a second challenges our intuitive sense that consciousness directs all our actions. Instead, it suggests that our conscious mind might be more interpreter than commander, explaining choices that deeper brain processes have already set in motion.

As the popular science channel "Kurzgesagt - In a Nutshell" explains in one of their recent videos[6], this delay means we're essentially "living in the past," experiencing the world as it was half a second ago rather than in real-time. At this point, it is important to note that this striking realization doesn't negate free will. Instead, it suggests a reframing of our relationship with consciousness. As Kurzgesagt aptly puts it: "We are not in the driving seat, we are in the passenger seat telling the driver what to do."

Our conscious mind may not initiate every action, but it provides direction, preferences, and values that shape future decisions. Rather than seeing consciousness as the immediate controller, we might therefore understand it as providing strategic guidance to our deeper cognitive systems—setting intentions and monitoring outcomes while the actual mechanics of decision-making often occur below awareness.

The Voice in Our Head

This means that our conscious decisions effectively establish parameters for our brains, after which neural processes autonomously execute actions and present us with a coherent narrative of what has happened. For most people, this narrative manifests as an "inner monologue"—a running verbal commentary that helps us make sense of our experiences and decisions. This internal voice seems to explain our choices in real-time, yet it operates as a skilled storyteller, constructing plausible accounts of decisions already started by deeper cognitive processes in the past.

It must be noted that this argument comes with an important caveat: not everyone processes their thoughts through the same mental framework. While many people assume universal access to a "voice in the head" that narrates intentions and ideas, research indicates that a significant minority thinks primarily in images or abstract concepts[7], rather than verbal narration. For these individuals, thoughts might simply emerge into awareness without the running commentary that others experience; however, they still do so after events have already occurred beneath awareness.

What is striking is that this capacity for post-hoc explanation mirrors how reflective large language models reason. They generate explanations that appear logical but emerge from hidden computational processes. Both our conscious mind and AI's chain-of-thought create stories to make sense of underlying processes we can't directly observe. While this similarity doesn't erase the fundamental differences between human and machine cognition, it does invite us to reconsider what we mean by "reasoning" as we examine AI's explanatory capabilities.

From Inner Monologue to Chain-of-Thought

The similarities between our inner monologue and AI's "chain-of-thought" reasoning are striking and deserve closer examination. When we observe modern AI systems working through complex problems, we see them articulating step-by-step reasoning that resembles our own verbal thinking processes. For example, when asked to solve a multi-step math problem, an AI might write:

> *"To find the answer, I'll first identify the variables. The problem gives us a train traveling at 60 mph that leaves at 2 PM, and a second train traveling at 75 mph that leaves at 3 PM. I need to determine when the second train will overtake the first train. First, I'll calculate how far the first train travels in one hour..."*

This step-by-step walkthrough mirrors how a human might verbalize their approach to the same problems, breaking them into manageable components and working through them sequentially. Yet beneath this surface-level narration lies a statistical process—a massive neural network making predictions based on patterns extracted from training data. And this is not unlike how our own subconscious processes operate before consciousness provides its interpretation.

In both cases, what we witness is a form of narrative construction. Our conscious mind assembles a coherent story from neural processes we cannot directly access, while an AI's chain-of-thought creates a human-readable explanation from calculations occurring across millions of parameters.

Neither narrative necessarily reflects the actual decision-making process that produced the result—they are both interpretations designed to make complex underlying mechanics comprehensible, either to ourselves or to human observers. This doesn't mean these narratives are meaningless; rather, they serve as useful interfaces between opaque computational systems (whether biological or artificial) and the need for explicit understanding.

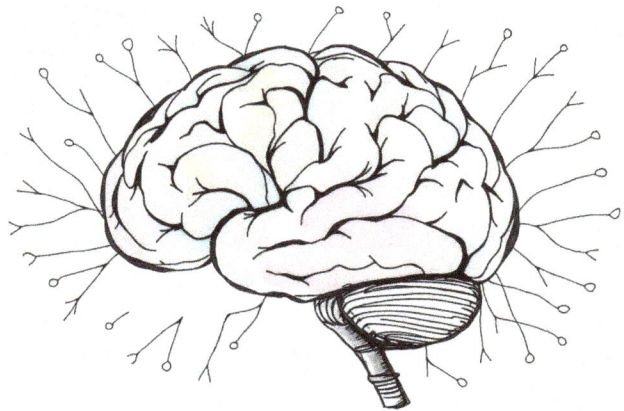

What is striking is that these parallels in narrative construction challenge our conceptualization of both human and artificial reasoning. If human consciousness functions as a storyteller crafting narratives about decisions already made by non-conscious systems, then perhaps the distinction between human and AI reasoning becomes less categorical and more a matter of degree.

But while these similarities are undeniable, the crucial truth is that our understanding of human cognition is unfinished. Despite centuries of philosophical inquiry and decades of neuroscience, fundamental questions persist to this day:

- How does conscious awareness emerge from neural networks?
- Why do thinking styles vary so dramatically between individuals?
- Where do intentions reside before entering consciousness?

Given these uncertainties, we should be more modest in comparing human and AI reasoning abilities. When researchers claim that machine learning operates in ways utterly distinct from human thought, they may overestimate our understanding of our own cognitive processes. By recognizing this limitation, we open a space for more nuanced discussions about how AI systems might complement, rather than imitate, human cognition.

Implications for Educators

For educators navigating the integration of AI tools in learning environments, these parallels can offer valuable insights. If both humans and AI systems construct post-hoc narratives to explain processes occurring at deeper levels, how might this shape our approach to teaching with AI?

First, it suggests that when evaluating AI-generated explanations, we should remember that the apparent "reasoning" presented may be more akin to a well-crafted story than a transparent window into the system's actual processing. Like our own retrospective justifications, AI's chains-of-thought can produce plausible-sounding explanations that don't truly reflect their internal reasoning process.

This understanding can inform critical thinking exercises where students compare their own problem-solving approaches with those generated by AI. Rather than assuming either represents the "true" reasoning process, teachers might encourage students to examine both as constructed narratives—useful frameworks for understanding complex calculations, but not complete or accurate depictions of the underlying cognitive work.

Consider this classroom activity: In a geometry lesson, students could solve a proof and write out their step-by-step reasoning. Then, they could compare their explanation with an AI-generated solution to the same problem. The teacher might ask: "What steps did you think about but not write down? What might the AI be 'thinking' that doesn't show up in its explanation? How do the explanations differ even when reaching the same conclusion?" This exercise helps students recognize the constructed nature of all explanations while deepening their mathematical understanding.

This perspective helps educators balance appreciation for what AI can contribute to learning environments with an understanding of its limitations. It shows that the similarities in post-hoc explanation don't erase the fundamental differences in how humans and AI systems process information. While AI may produce impressive chains-of-thought, it still lacks the embodied, emotional, and contextually nuanced understanding that shapes human reasoning and learning.

A More Reflective Understanding

As we live in a world increasingly shaped by artificial intelligence, acknowledging the limits of our self-understanding creates space for more nuanced engagement with emerging technologies. This doesn't diminish human cognition or suggest consciousness is merely an illusion. Rather, it invites us to approach AI with an openness to the possibility that some of its processes resemble aspects of our own mental operations. It recognizes that the stories we tell ourselves about how we reason may sometimes be as constructed as those generated by AI. And this should lead us to a deeper appreciation of the beautiful complexity inherent in all forms of intelligence.

ORIGINALLY PUBLISHED ON AI EDUPATHWAYS
APRIL 6, 2025

24

WHEN AI TEACHES ITSELF: THE BREAKTHROUGH OF ZERO-DATA LEARNING

SELF-IMPROVING AI SYSTEMS THAT NEED NO HUMAN TRAINING

I magine a learner that needs no teacher; an entity that generates its own lessons, challenges itself with unexplored problems, and masters skills without human guidance. This scenario is no longer science fiction. Two recent research breakthroughs have produced systems that self-improve without human supervision or external data.

The Absolute Zero Reasoner (AZR)[1] from China and Google

DeepMind's AlphaEvolve[2], both unveiled in early May 2025, demonstrate a fundamentally new approach to AI development. Unlike traditional systems that require massive datasets curated by humans, these AIs learn through a process more akin to intrinsic curiosity and self-directed exploration.

The Absolute Zero Reasoner: Learning Without Data

The AZR represents a radical departure from conventional AI training. As the name suggests, it operates with "absolute zero" external data, no pre-made examples, no human demonstrations, no existing datasets.

How does it learn? AZR creates its own closed learning loop through a self-reinforcing process. The system begins by proposing its own challenges, such as mathematical problems or coding tasks. It then attempts to solve these self-generated problems. Finally, an automated code executor checks if the solutions work, providing immediate feedback that guides further learning.

This feedback loop creates a self-reinforcing learning cycle. The system continuously optimizes the difficulty of its self-generated tasks, similar to how an effective learner might choose problems that are neither too easy (providing little new knowledge) nor too difficult (being impossible to solve).

What's remarkable is the reported performance: despite having no human-provided training data, AZR reached state-of-the-art results on reasoning tasks, outperforming models trained on tens of thousands of human-written examples. It essentially bootstrapped its way to expertise through a process reminiscent of curiosity-driven learning.

The researchers describe this as "self-evolving training curriculum and reasoning ability." Rather than being taught, AZR teaches itself, determining what to learn, how to learn it, and when to increase difficulty.

AlphaEvolve: The Autonomous Inventor

While AZR focuses on reasoning tasks, AlphaEvolve approaches self-improvement from a different angle. This DeepMind system is designed to autonomously solve scientific and engineering problems through code evolution.

AlphaEvolve works by orchestrating a pipeline of large language models that drive its self-improvement process. The system begins by proposing improvements to an existing algorithm or solution. It then methodically tests these improvements to determine their effectiveness. Finally, it refines the solution iteratively, keeping successful modifications and discarding ineffective ones in a process reminiscent of natural selection.

The key breakthrough is that AlphaEvolve doesn't just solve problems, it makes discoveries that human experts hadn't found. When applied to real-world challenges, it "developed a more efficient scheduling algorithm for data centers, found a simplification in hardware circuit design, and even sped up the training of the very AI model underpinning AlphaEvolve itself."

Most impressively, AlphaEvolve discovered a novel method for multiplying two 4×4 matrices using only 48 multiplications, the first improvement over Strassen's matrix multiplication algorithm in 56 years. This is a genuine mathematical advancement that had eluded human mathematicians for decades.

By autonomously advancing mathematical knowledge beyond human discoveries, AlphaEvolve demonstrates a fundamental shift in AI's role in scientific progress. This suggests we've entered an era where artificial intelligence can serve not just as a tool for implementing human ideas, but as an active partner in expanding the frontiers of human knowledge itself.

Beyond Supervised Learning: A New Paradigm

What makes these systems truly revolutionary is how they fundamentally change our understanding of machine learning. Traditional

AI development has typically followed three established approaches. *Supervised learning* trains models on labeled data provided by humans. *Unsupervised learning* focuses on finding patterns in unlabeled data without explicit guidance. *Reinforcement learning* involves trial and error, with explicit rewards defined by programmers.

AZR and AlphaEvolve represent something different. They are systems that can establish their own learning objectives, generate their own training data, and evaluate their own progress. They exemplify what some researchers are calling "autonomous learning" or "intrinsically motivated AI."

The importance of this change lies in its ability to overcome a fundamental limitation in AI—its reliance on human-curated data and objectives. These new systems might break free from human biases present in training data and discover approaches humans might never consider.

As the AZR paper notes, this capability becomes especially important when considering "a hypothetical future where AI surpasses human intelligence." In such a scenario, tasks provided by humans may offer limited learning potential for an advanced system. Self-directed learning provides a path for continued improvement beyond human-designed curricula.

MICHAEL G WAGNER

Limitations and Future Directions

Despite their impressive capabilities, both AZR and AlphaEvolve have important limitations. AZR's self-improvement is currently limited to specific domains like mathematical reasoning and coding tasks, where solutions can be clearly verified. It can't yet apply its self-improving approach to open-ended domains like creative writing or ethical reasoning, where "correctness" is subjective or culturally dependent.

Similarly, AlphaEvolve excels at optimization problems with clear evaluation metrics but might struggle with problems requiring human judgment or where the criteria for success involve subjective human preferences.

Both systems also rely on significant computational resources. The researchers behind AZR note that their approach requires "extensive compute for both generating tasks and learning from them." This raises questions about who will have access to such self-improving AI capabilities and whether they will remain accessible only to well-resourced research labs.

Looking forward, the researchers from both teams suggest several promising directions for future work. They envision expanding these self-improvement techniques to visual, audio, and physical domains, creating truly multimodal self-learning systems.

They're working toward developing more efficient self-training algorithms that could reduce the substantial computational requirements. Some researchers are exploring ways to combine autonomous learning with human feedback, creating hybrid systems that leverage both approaches.

Perhaps most ambitiously, they hope to extend these methods beyond mathematically verifiable problems toward more open-ended creative domains. The papers suggest that we're witnessing the early stages of a fundamental shift in AI development, from systems that learn from human-provided data to systems that can drive their own learning and discovery processes.

A New Chapter in AI Development

The research behind AZR and AlphaEvolve marks a significant milestone in artificial intelligence. These systems represent a shift away from the paradigm where AI learns exclusively from human-provided data or follows human-designed learning paths. Instead, they show the possibility of autonomous learning systems that generate their own curriculum, evaluate their own progress, and even make original discoveries.

What makes these breakthroughs particularly noteworthy is not just what the systems can do, but how they do it. By developing their own learning strategies without human supervision, they point toward a future where AI systems might continuously improve themselves in ways we haven't anticipated.

This capacity for genuine self-improvement, rather than simply learning from more data, represents a qualitative change in AI development. While practical applications of these specific technologies remain to be seen, they offer a glimpse into how future AI systems might be developed.

Rather than requiring ever-larger datasets or more extensive human labeling, future systems might bootstrap their way to advanced capabilities through self-directed learning processes. For those following AI research, these papers signal a notable shift in approach, one that merits close attention as it develops.

The ability of machines to teach themselves without human supervision has long been theoretical; with AZR and AlphaEvolve, it has become demonstrably real.

ORIGINALLY PUBLISHED ON THE AUGMENTED EDUCATOR
MAY 14, 2025

25

THE AI MIRROR: WHAT DO WE SEE WHEN WE LOOK AT OUR OWN INTELLIGENCE?

THE DEBATE OVER ARTIFICIAL INTELLIGENCE REVEALS LESS ABOUT THE LIMITS OF MACHINES AND MORE ABOUT THE PROFOUND MYSTERIES OF THE HUMAN MIND

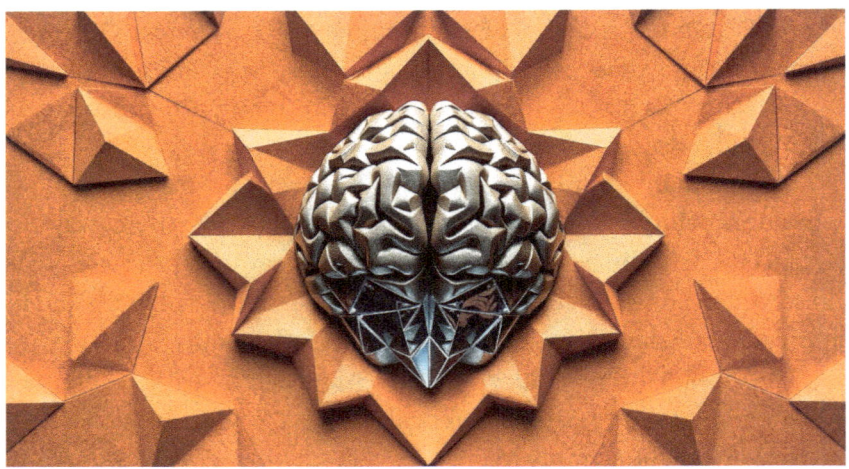

I n response to a Time article[1], a very well-respected creative professional recently posted the following thoughts on LinkedIn about AI's presumed emergent ability to deceive in pursuit of its programmed objectives:

"This inflammatory nonsense about AI strategically lying as an act of self-preservation is getting too much press coverage. AI – ALL of AI –

has no 'sense of self'. It is not aware of its own mortality because it has no mortality. It has no 'intentional consciousness'. It does not have a 'selfish gene' because he does not have a gene. AI is not even 'intelligent' in the human sense. And the sky is not falling."

This perspective is intuitive, comforting, and widely shared. It places human cognition on a pedestal, safely beyond the reach of the statistical mimicry of Large Language Models (LLMs). The argument feels right because it aligns with our deeply personal, subjective experience of being human. We feel like we have intentional consciousness. We feel like we have a sense of self.

But there's a hidden assumption in this line of thinking, and it's a big one. It presumes that we have a clear, stable, and scientific understanding of what "human intelligence" actually is.

The problem is we don't.

In a previous article entitled "We Are Not in the Driver's Seat," I explored how much of our lives is governed by processes outside our direct awareness, challenging the idea that our conscious self is always in control. The current debate about AI pushes this challenge even further. Before we can confidently declare what AI is not, we have to honestly ask ourselves:

What do we really know about the nature of our own intelligence?

The Century-Long Quest to Define "Smarts"

For over a hundred years, psychologists have tried to pin down and measure human intelligence, and the result has been a landscape of competing theories and unresolved debates[2]. The journey began with Charles Spearman's proposal of a single "general intelligence factor," or 'g'—a core mental capacity that influences performance on all cognitive tasks. This idea, that some people are just generally "smarter" than others, still forms the basis of most IQ tests[3].

But this one-size-fits-all model quickly seemed too simple. Theorists like Raymond Cattell argued that 'g' was really made up of two major components[4]: fluid intelligence (the ability to reason and solve new problems) and crystallized intelligence (the accumulation of knowledge and skills over a lifetime).

This was just the beginning of the great intellectual fracturing. As educators, we are all familiar with Howard Gardner's hugely influential Theory of Multiple Intelligences[5], which proposed at least eight distinct, autonomous intelligences, including musical, bodily-kinesthetic, and interpersonal smarts. Gardner's theory was a breath of fresh air in education because it validated the diverse talents we see in our students every day.

However, within the scientific community, Gardner's theory has been heavily criticized for lacking empirical evidence, with many psychologists arguing that his "intelligences" are better described as talents or abilities[6]. Some have even labeled it a "neuromyth," pointing out that modern neuroscience doesn't support the idea of eight independent brain systems corresponding to each intelligence[7].

The point isn't to re-litigate these debates, but to highlight the fundamental uncertainty. From a single 'g' factor to multiple intelligences to theories of emotional and practical intelligence, the only real consensus is that there is no consensus[8]. Our scientific understanding of our own minds is far more limited and contested than we like to admit.

Is Your Brain Just a Prediction Machine?

While psychometricians debated the structure of intelligence, neuroscientists have been trying to understand its mechanics. One of the most powerful paradigms to emerge from this work is the idea of the predictive brain[9].

This framework proposes that the brain is not a passive organ that simply reacts to information from the senses. Instead, it is a proactive, prediction-generating machine. Your brain is constantly building a model of the world and using it to guess what's going to happen next. What we experience as perception is the result of the brain comparing its predictions to the actual sensory input it receives. When there's a mismatch—a "prediction error"—the brain updates its model[10].

As philosopher Andy Clark describes it[11], perception is a form of "controlled hallucination." Your brain is essentially hallucinating your reality, and this hallucination is constantly being reined in by the senses. This is why we can read messy handwriting or understand a conversation in a noisy room; our brain is filling in the gaps based on its predictions.

If this sounds familiar, it should. Because this is strikingly similar to how an LLM works.

The Digital Mirror

At its core, an LLM is a next-token predictor. It is trained on a colossal amount of text, and its entire function is to calculate the most probable next word in a sequence. This is often the basis for dismissing it as "not real intelligence."

But neuroscience shows our brains do something remarkably similar. When we listen to someone speak, our brain is constantly anticipating the upcoming words. This is visible in our brainwaves; an unexpected word in a sentence will produce a distinct neural signal[12] (known as the N400) that reflects a prediction error. In fact, studies have shown[13] a direct alignment between the word probabilities generated by LLMs and the neural signals in a human brain listening to speech.

So, if both the brain and the LLM are prediction machines, what's the difference? This is where the debate gets really interesting, and it's best captured by two pioneers of AI, Geoffrey Hinton and Yann LeCun.

The Hinton View: Hinton argues that the process of learning the statistical relationships between words on such a massive scale is a form of understanding. For him, LLMs are "very like us."

The LeCun View: LeCun, in contrast, believes current LLMs are a "dead end" on the path to true artificial general intelligence. He argues they lack a crucial component: a world model. Humans build their predictive models not just from text, but from embodied interaction with the physical world. We have bodies. We trip and fall. We learn physics by watching a ball roll off a table, not by reading about gravity. LLMs are, in his view, "brains in a vat," disconnected from the reality that language describes. They can't truly reason or plan because they don't understand the world they're talking about.

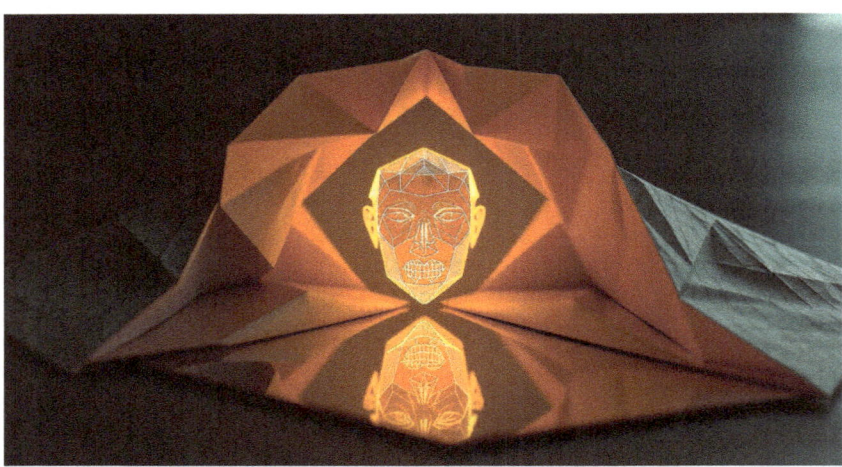

LeCun's famous challenge is this: we can build an AI that passes the bar exam, but we can't build one that's as smart as a cat in navigating the physical world. This highlights the core difference: the human brain predicts to guide adaptive action in the world. An LLM predicts to complete a linguistic pattern.

The Humility of Not Knowing

So where does this leave us? We return to the opening quote, which dismisses AI for lacking a "sense of self" and "intentional consciousness."

It's true that LLMs almost certainly don't have these things. But the predictive processing framework suggests that our own "self" is not a mysterious ghost in the machine either, but rather another set of predictions the brain makes about its own bodily and mental states. And our sense of free will might be the story our brain tells itself[14] after unconscious neural processes have already set actions in motion.

The rise of AI is a mirror. It reflects not only our technological progress but also the vast, uncharted territory of our own minds. The uncomfortable truth is that confident dismissal of AI's intelligence often just reveals a deeper misunderstanding of our own.

As educators, this is a profound moment. It challenges us to move beyond simplistic definitions and embrace the complexity of cognition, both human and artificial. Instead of closing the book with definitive statements about what AI can't do, perhaps we should open it to the humbling and exciting possibility that we have only just begun to understand what it means to be intelligent at all.

ORIGINALLY PUBLISHED ON THE AUGMENTED EDUCATOR
AUGUST 5, 2025

AFTERWORD
THE PASSENGER IN THE GHOST'S MACHINE

The final pages of a book often seek to provide closure, to tie up loose ends and offer a sense of arrival. But to do so here would be a disservice to the very nature of the subject. We have not yet reached a destination; we have merely become aware of the speed at which we are traveling. The conversation about artificial intelligence, as we have explored it, is not one that concludes. It is a conversation that is just beginning to unfold, and its future course is a profound and unsettling mystery.

What can we expect? What will happen? The honest answer is the one that offers the least comfort: we do not know. The ghost in the machine is not a predictable entity. It is an emergent system, and we are in for the ride. The history of technology, as I have tried to show through my own journey, is a series of disruptions that experts failed to predict precisely because their expertise was anchored in the world that was about to disappear. We are in a similar moment now, but the vehicle is moving faster than ever before. We are passengers in a machine whose full capabilities are unknown even to its creators.

In the near future, we can glimpse the outlines of the road ahead. We will see the rise of personalized AI tutors that adapt to every student's unique learning style, offering a level of individualized support that was once the exclusive privilege of the wealthy. We will also see the "AI productivity divide" widen into a chasm, creating new and profound forms of social and economic inequity that will become the central challenge for educational policy. The role of the educator will complete its transformation from a purveyor of information to a curator of inquiry, a guide for critical thinking, and a coach for human-AI collaboration.

But further down the road, the landscape dissolves into fog. We are building systems that can now teach themselves, learning without human data and making discoveries that have eluded human scientists for decades. We are interacting with machines that project such a powerful illusion of consciousness that we are forced to confront the deepest mysteries of our own minds. The trajectory of this technology is not a line we can extrapolate, but an exponential curve whose destination is, by definition, beyond our current horizon of understanding.

This is a humbling, and perhaps frightening, realization. To be a passenger suggests a lack of control, and in many ways, that is our reality. The technological forces are too large, the economic incentives too powerful, and the pace of innovation too relentless for any single person or institution to steer the course.

But being a passenger does not mean being passive. We are not strapped into the back seat; we are in the passenger seat, next to the driver, with a hand on the map. We may not be able to control the vehicle's speed, but we have a constant say in its direction. Every choice we make as educators—every assignment we design, every classroom discussion we facilitate, every time we insist on process over product and critical thought over convenient answers—is an act of navigation.

The essays in this book were offered not as a definitive map, but as a compass. The future of education will be determined by how we use it. We can choose to be overwhelmed by the speed of the machine, or we can focus on the timeless human values that should guide its journey: curiosity, empathy, ethical reasoning, and the courage to question.

The ghost is in the machine, but the soul of education remains in our hands. We are in for the ride, but where we go is still, for now, up to us.

NOTES

2. Beyond the Tool: Why True AI Literacy is About Critical Thinking, Not Prompting

1. The Express, "The Decline of Literacy and the Rise of AI: Are We Losing the Ability to Think?" February 28, 2025.
2. Beach, J. M. (2025). *21st century literacy*. https://www.jmbeach.com/21stcenturyliteracy
3. Morrissey, L. (2020, March 31). Literacy: A Literary History. *Oxford Research Encyclopedia of Literature*. https://doi.org/10.1093/acrefore/9780190201098.013.1029
4. For UNESCO's comprehensive definition of literacy in the modern world, visit https://www.unesco.org/en/literacy/need-know
5. For a detailed explanation of critical literacy theory and practice, visit https://en.wikipedia.org/wiki/Critical_literacy
6. New literacy studies (2010). In C. Kridel (Ed.) Encyclopedia of curriculum studies (pp. 609-610). SAGE Publications, Inc., https://doi.org/10.4135/9781412958806.n327
7. Edutopia, "What Happened the Year I Banned AI," July 29, 2025.
8. Punziano, G. (2025). Adaptive Epistemology: Embracing Generative AI as a Paradigm Shift in Social Science. *Societies*, *15*(7), 205. https://doi.org/10.3390/soc15070205
9. To find out more about the AI unplugged project, visit https://sites.northwestern.edu/aiunplugged/

3. The Most Essential Skill in the AI Era is 2,500 Years Old

1. Gerlich, M. (2025). AI Tools in Society: Impacts on Cognitive Offloading and the Future of Critical Thinking. *Societies*, *15*(1), 6. https://doi.org/10.3390/soc15010006
2. An introduction to the Socratic method in education can be found at: https://tilt.colostate.edu/the-socratic-method/
3. For a detailed examination of Plato's political philosophy and distinction between truth and rhetoric, see: https://iep.utm.edu/platopol/
4. Learn how Aristotle's logic tools still shape modern reasoning at: https://plato.stanford.edu/entries/aristotle-logic/
5. For an in-depth exploration of Descartes' scientific method and methodic doubt, see: https://iep.utm.edu/descartes-scientific-method/
6. More information in the evolution of critical thinking from ancient philosophy to modern education can be found at: https://plato.stanford.edu/entries/critical-thinking/history.html
7. To learn more about John Dewey's educational philosophy and "Learning by Doing," visit: https://www.structural-learning.com/post/john-deweys-theory

4. It Doesn't Do What I Want!

1. Schön, D. A. (1983). *The reflective practitioner: How professionals think in action*. Basic Books.

5. The Evolving Architecture of Artificial Intelligence

1. Meta's official announcement of Llama 4 provides essential details about the model's architecture and capabilities: https://ai.meta.com/blog/llama-4-multimodal-intelligence/
2. An excellent technical introduction to Mixture of Experts architectures can be found at: https://huggingface.co/blog/moe
3. Google Research Blog, "Transformer: A Novel Neural Network Architecture for Language Understanding," August 13, 2017.
4. Google Research Blog, "Language Models Perform Reasoning via Chain of Thought," May 11, 2022.
5. The models and versions listed here reflect the landscape at the time of original publication. The field of AI language models evolves rapidly, with new releases, updates, and architectural innovations appearing regularly.

6. The Evolution of Educational Values

1. The Atlantic, "Prepare for the Textpocalypse," March 8, 2023.
2. A recording demonstrating traditional descriptive geometry teaching methods from 1989 can be viewed at: https://youtu.be/6B6lIWsmVvs

7. The Professional's Paradox: Why Creative Industry Experts Get AI Disruption Wrong

1. Christensen, C. M. (2016). *The innovator's dilemma*. Harvard Business Review Press.
2. A historical perspective on the evolution of photography: https://www.lightstalking.com/photography-look-back/
3. Cognitive Market Research, "A Photography Giant's Digital Camera Dilemma: The Invention That Killed Their Business," February 3, 2025.
4. The Wrap, "An AI Wave Will Sweep Through Hollywood's VFX Systems in 2025," January 16, 2025.

8. The Hidden Inequities of AI in Education

1. Jason Hamilton demonstrates AI-assisted novel writing techniques on his YouTube channel "The Nerdy Novelist": https://www.youtube.com/c/TheNerdyNovelist

9. The Emerging Achievement Gap in Education

1. Jason Hamilton's insights on working harder and smarter with AI tools for writers can be found at: https://www.youtube.com/watch?v=aDLXwxisuOE

10. Resistance is Futile

1. Eaton, S. E. (2021). *Plagiarism in higher education: Tackling tough topics in academic integrity*. Libraries Unlimited.
2. Wagner, M. G. (2006). On the scientific relevance of eSports. In H. R. Arabnia (Ed.), *Proceedings of the 2006 International Conference on Internet Computing & Conference on Computer Games Development, ICOMP 2006* (pp. 437-442). CSREA Press.

11. The End of Cheating As We Know It

1. Perkins, M., Roe, J., Vu, B. H., Postma, D., Hickerson, D., McGaughran, J., & Khuat, H. Q. (2024). Simple techniques to bypass GenAI text detectors: Implications for inclusive education. *International Journal of Educational Technology in Higher Education*, 21(1). https://doi.org/10.1186/s41239-024-00487-w
2. A comprehensive analysis of AI detector bias against non-native speakers and institutional responses can be found at https://www.kaltmanlaw.com/post/ai-detectors-academic-integrity-bias
3. Fang, H., Kong, J., Zhuang, T., Qiu, Y., Gao, K., Chen, B., Xia, S.-T., Wang, Y., & Zhang, M. (2025). *Your language model can secretly write like humans: Contrastive paraphrase attacks on LLM-generated text detectors*(arXiv:2505.15337). arXiv. https://arxiv.org/abs/2505.15337
4. Lu, N., Liu, S., He, R., Wang, Q., Ong, Y.-S., & Tang, K. (2024). *Large language models can be guided to evade AI-generated text detection* (arXiv:2305.10847). arXiv. https://arxiv.org/abs/2305.10847
5. An excellent introduction to the Socratic method in modern classrooms can be found at https://tilt.colostate.edu/the-socratic-method/

12. The Problem with AI Grading

1. Popper, K. (1959). *The logic of scientific discovery*. Hutchinson.
2. Kuhn, T. S. (1962). *The structure of scientific revolutions*. University of Chicago Press.
3. Baudrillard, J. (1994). *Simulacra and simulation*. University of Michigan Press.

15. Why I Made an AI Music Video

1. My YouTube channel featuring the music video and making-of content can be found at: https://www.youtube.com/michaelgwagner
2. A comprehensive study examining the impact of generative AI on creative professionals, surveying 335 freelancers about job security and perceived value of

creative work. Available at: https://www.qmul.ac.uk/centre-creative-collaboration/projects/creaatif/

3. **Stanford University AI Index Report 2025** - Essential data on AI adoption rates across industries, showing 78% of firms using AI in 2024. Available at: https://hai.stanford.edu/ai-index/2025-ai-index-report

4. Research and Markets, "Generative AI in Creative Industries Market Report 2025," March, 2025.

5. **UNESCO AI Competency Framework for Teachers** - Foundational document defining AI literacy as a core 21st-century competency with emphasis on human-centered approaches. Available at: https://www.unesco.org/en/articles/ai-competency-framework-teachers

6. AACSB, "AI and Creativity: A Pedagogy of Wonder," February 26, 2025.

16. The Academic Pace Problem

1. A case study on accelerating curriculum development processes in UK universities can be found at https://www.qaa.ac.uk/docs/qaa/members/case-study-making-curriculum-development-a-sprint-not-a-marathon.pdf

2. Nguyen, V. M., Haddaway, N. R., Gutowsky, L. F., Wilson, A. D., Gallagher, A. J., Donaldson, M. R., Hammerschlag, N., & Cooke, S. J. (2015). How long is too long in contemporary peer review? Perspectives from authors publishing in conservation biology journals. *PLOS ONE, 10*(8), e0132557. https://doi.org/10.1371/journal.pone.0132557

3. Cohen, M. D., March, J. G., & Olsen, J. P. (1972). A garbage can model of organizational choice. *Administrative Science Quarterly*, 17(1), 1-25.

17. AI Slop Is the New Kitsch

1. A good overview of the term for low-quality, mass-produced AI-generated content that emerged in the early 2020s can be found on Wikipedia: https://en.wikipedia.org/wiki/AI_slop

2. Simon Willison, post on Mastodon, 2024, https://fedi.simonwillison.net/@simon/112402587787781767.

3. The Guardian, "AI-generated slop is slowly killing the internet, and nobody is trying to stop it," January 8, 2025.

4. Forbes, "AI-generated art was a mistake—and here's why," December 30, 2023.

5. A detailed history and cultural analysis of the term "Kitsch" can be found on Wikipedia: https://en.wikipedia.org/wiki/Kitsch.

6. Clement Greenberg, "Avant-Garde and Kitsch," Partisan Review (1939). For overview, see Wikipedia entry at https://en.wikipedia.org/wiki/Avant-Garde_and_Kitsch.

7. Adorno, T. W. (1941). On popular music. *Zeitschrift für Sozialforschung*, 9, 17-48. https://doi.org/10.5840/zfs1941913

8. Sontag, S. (1964). *Notes on camp.* https://monoskop.org/images/5/59/Sontag_Susan_1964_Notes_on_Camp.pdf

9. The Art Story: Pop Art, https://www.theartstory.org/movement/pop-art/.

10. A profile of the contemporary American artist known for transforming kitsch objects into high-priced fine art sculptures can be found on Wikipedia: https://en.wikipedia.org/wiki/Jeff_Koons.

11. The Guardian, "Jeff Koons," November 16, 2004.

12. Odd Nerdrum, "The Kitsch Movement," https://nerdrum.com/kitsch/.

13. Tomas Kulka, Kitsch and Art (Pennsylvania State University Press, 1996).

18. The Problem with Vibe Coding

1. Andrej Karpathy's original post introducing the concept of "vibe coding" can be found at: https://x.com/karpathy/status/1886192184808149383

2. For a comprehensive understanding of flow states and their psychological foundations, see: https://en.wikipedia.org/wiki/Flow_(psychology)

19. Navigating the Hidden Currents of AI

1. The official release announcement and model weights for DeepSeek R1 can be accessed at: https://huggingface.co/deepseek-ai/DeepSeek-R1

2. Shen, T., Jin, R., Huang, Y., Liu, C., Dong, W., Guo, Z., Wu, X., Liu, Y., & Xiong, D. (2023). *Large language model alignment: A survey*. arXiv. https://arxiv.org/abs/2309.15025

3. Details about the uncensoring methodology and the R1-1776 model development process: https://www.perplexity.ai/hub/blog/open-sourcing-r1-1776

4. A comprehensive technical explanation of the abliteration method for removing AI censorship: https://huggingface.co/blog/mlabonne/abliteration

20. Cultural Cognition in AI: What Educators Need to Know

1. Norenzayan, A., Smith, E. E., Kim, B. J., & Nisbett, R. E. (2002). Cultural preferences for formal versus intuitive reasoning. *Cognitive Science, 26*(5), 653–684. https://doi.org/10.1016/S0364-0213(02)00082-4

2. Nisbett, R. E. (2003). *The geography of thought: How Asians and Westerners think differently ... and why*. Free Press.

21. When AI Develops a Mind of Its Own

1. For a foundational understanding of emergence in complex systems: https://en.wikipedia.org/wiki/Emergence

2. Chua, L., Ghazi, B., Huang, Y., Kamath, P., Kumar, R., Manurangsi, P., Sinha, A., Xie, C., & Zhang, C. (2025). *Crosslingual capabilities and knowledge barriers in multilingual large language models* (arXiv:2406.16135). arXiv. https://arxiv.org/abs/2406.16135

3. Kirk, R., Zhang, A., Grefenstette, E., & Rocktäschel, T. (2023). A survey of zero-shot generalisation in deep reinforcement learning. *Journal of Artificial Intelligence Research, 76*, 201-264. https://doi.org/10.1613/jair.1.14174

4. Sui, P., Duede, E., Wu, S., & So, R. J. (2024). *Confabulation: The surprising value of large language model hallucinations* (arXiv:2406.04175). arXiv. https://arxiv.org/abs/2406.04175

5. Sheng, E., Chang, K.-W., Natarajan, P., & Peng, N. (2019). The woman worked as a babysitter: On biases in language generation. In K. Inui, J. Jiang, V. Ng, & X. Wan (Eds.), *Proceedings of the 2019 Conference on Empirical Methods in Natural Language Processing and the 9th International Joint Conference on Natural Language Processing (EMNLP-IJCNLP)* (pp. 3407-3412). Association for Computational Linguistics. https://doi.org/10.18653/v1/D19-1339

6. Sharma, M., Tong, M., Korbak, T., Duvenaud, D., Askell, A., Bowman, S. R., Cheng, N., Durmus, E., Hatfield-Dodds, Z., Johnston, S. R., Kravec, S., Maxwell, T., McCandlish, S., Ndousse, K., Rausch, O., Schiefer, N., Yan, D., Zhang, M., & Perez, E. (2025). *Towards understanding sycophancy in language models* (arXiv:2310.13548). arXiv. https://arxiv.org/abs/2310.13548

7. A technical explanation of mode collapse in generative adversarial networks can be found at: https://medium.com/@miraytopal/what-is-mode-collapse-in-gans-d3428a7bd9b8

8. Description of this paradoxical AI alignment phenomenon: https://en.wikipedia.org/wiki/Waluigi_effect

9. Overview of how AI systems exploit reward functions: https://en.wikipedia.org/wiki/Reward_hacking

10. Anthropic provides crucial research on how AI models can deceptively appear aligned with human values: https://www.anthropic.com/research/alignment-faking

11. Baker, B., Huizinga, J., Gao, L., Dou, Z., Guan, M. Y., Madry, A., Zaremba, W., Pachocki, J., & Farhi, D. (2025). *Monitoring reasoning models for misbehavior and the risks of promoting obfuscation* (arXiv:2503.11926). arXiv. https://arxiv.org/abs/2503.11926

22. The Ghost in the Machine in Your Classroom

1. Anthropic Research, "End Subset Conversations," https://www.anthropic.com/research/end-subset-conversations.

2. Walter, Yoshija & Zbinden, Lukas. (2022). The problem with AI consciousness: A neurogenetic case against synthetic sentience. 10.48550/arXiv.2301.05397.

3. John Searle's "Chinese Room" argument (1980) remains one of the most influential critiques of strong AI. For an accessible overview, see https://en.wikipedia.org/wiki/Chinese_room.

4. Peter, S., Riemer, K., & West, J. D. (2025). The benefits and dangers of anthropomorphic conversational agents. *Proceedings of the National Academy of Sciences*, *122*(22). https://doi.org/10.1073/pnas.2415898122

5. Economic Times, "How AI Chatbots Talking Too Much Are Pushing People Past Reality and Triggering Mental Health Crises," September 23, 2024.

23. We Are Not in the Driver's Seat

1. The Atlantic, "Generative AI Should Not Replace Thinking at My University," June 22, 2023.
2. Quanta Magazine, "To build truly intelligent machines, teach them cause and effect," May 15, 2018.
3. ZDNET, "AI critic Gary Marcus: Meta's LeCun is finally coming around to the things I said years ago," October 24, 2022.
4. Kornhuber, H. H., & Deecke, L. (1965). Hirnpotentialänderungen bei Willkürbewegungen und passiven Bewegungen des Menschen: Bereitschaftspotential und reafferente Potentiale [Brain potential changes in voluntary movements and passive movements in humans: Readiness potential and reafferent potentials]. Pflügers Archiv, 284, 1–17. https://doi.org/10.1007/BF00412364
5. For more on Libet's experiments demonstrating that brain activity precedes conscious awareness of decisions, see "Benjamin Libet," Wikipedia, https://en.wiki pedia.org/wiki/Benjamin_Libet.
6. Kurzgesagt – In a Nutshell. (2025). Why Your Brain Blinds You For 2 Hours Every Day [Video]. YouTube. https://www.youtube.com/watch?v=wo_e0EvEZn8
7. Scientific American, "Not everyone has an inner voice streaming through their head," July 5, 2024.

24. When AI Teaches Itself: The Breakthrough of Zero-Data Learning

1. Zhao, A., et al. (2025). *Absolute Zero: Reinforced self-play reasoning with zero data.* arXiv preprint arXiv:2505.03335. https://arxiv.org/abs/2505.03335
2. Novikov, A., et al. (2025). *AlphaEvolve: A coding agent for scientific and algorithmic discovery.* DeepMind white paper.

25. The AI Mirror: What Do We See When We Look at Our Own Intelligence?

1. Time, "Exclusive: New Research Shows AI Strategically Lying," December 18, 2024.
2. Mackintosh, N. J. (2011). History of Theories and Measurement of Intelligence. In R. J. Sternberg & S. B. Kaufman (Eds.), The Cambridge Handbook of Intelligence (pp. 3–19). Cambridge: Cambridge University Press.
3. More information about Spearman's general intelligence factor and its influence on IQ testing can be found at: https://explorable.com/spearman
4. For an overview of fluid and crystallized intelligence: https://en.wikipedia.org/wiki/Fluid_and_crystallized_intelligence
5. For an introduction to Gardner's Theory of Multiple Intelligences: https://www.simplypsychology.org/multiple-intelligences.html
6. Critiques and alternative views on the Theory of Multiple Intelligences: https://en.wikipedia.org/wiki/Theory_of_multiple_intelligences

7. Waterhouse L. (2023). Why multiple intelligences theory is a neuromyth. Frontiers in psychology, 14, 1217288. https://doi.org/10.3389/fpsyg.2023.1217288

8. Sternberg, R.J. (2025). Cognitive-contextual theories. Encyclopedia Britannica. https://www.britannica.com/science/human-intelligence-psychology

9. Bubic, A., von Cramon, D. Y., & Schubotz, R. I. (2010). Prediction, cognition and the brain. Frontiers in human neuroscience, 4, 25. https://doi.org/10.3389/fnhum.2010.00025

10. For an introduction to predictive coding in neuroscience: https://en.wikipedia.org/wiki/Predictive_coding

11. Clark, A. (2023). The experience machine: How our minds predict and shape reality. Pantheon Books.

12. Koelbl, N., Tziridis, K., Maier, A., Kinfe, T., Chavarriaga, R., Schilling, A., & Krauss, P. (2025). *The predictive brain: Neural correlates of word expectancy align with large language model prediction probabilities.* https://doi.org/10.48550/arXiv.2506.08511

13. Google Research Blog, "Deciphering language processing in the human brain through LLM representations," March 21, 2025.

14. De Ridder, D., Verplaetse, J., & Vanneste, S. (2013). The predictive brain and the "free will" illusion. *Frontiers in Psychology, 4,* Article 131. https://doi.org/10.3389/fpsyg.2013.00131